T0231251

GROUNDWATER REMEDIATION

WATER QUALITY MANAGEMENT LIBRARY
VOLUME 8

GROUNDWATER REMEDIATION

EDITED BY

Randall J. Charbeneau, Ph.D., P.E.
Philip B. Bedient, Ph.D., P.E.
Raymond C. Loehr, Ph.D., P.E.

LIBRARY EDITORS

W. W. ECKENFELDER, D.Sc., P.E.   J. F. MALINA, JR., Ph.D., P.E., D.E.E.   J. W. PATTERSON, Ph.D.

CRC Press
Taylor & Francis Group
Boca Raton   London   New York

CRC Press is an imprint of the
Taylor & Francis Group, an informa business

CRC Press
6000 Broken Sound Parkway, NW
Suite 300, Boca Raton, FL 33487
270 Madison Avenue
New York, NY 10016
2 Park Square, Milton Park
Abingdon, Oxon OX14 4RN, UK

Water Quality Management Library—Volume 8
a TECHNOMIC®publication

Published in the Western Hemisphere by
Technomic Publishing Company, Inc.
851 New Holland Avenue
Box 3535
Lancaster, Pennsylvania 17604 U.S.A.

Distributed in the Rest of the World by
Technomic Publishing AG

10 9 8 7 6 5 4 3

Main entry under title:
  Water Quality Management Library—Volume 8 / Groundwater Remediation

A Technomic Publishing Company book
Bibliography: p.
Includes index p. 185

Library of Congress Card No. 92-53524
ISBN No. 87762-943-9 (Volume 8)
ISBN No. 87762-944-7 (8-Volume Set)

Table of Contents

Foreword

THE first years of the final decade of the twentieth century have seen dramatic changes in the political structure of the world, especially in Eastern Europe and the countries which constituted the former Union of Soviet Socialist Republics. Equally dramatic have been international upheavals in the field of environmental protection. These range from the uncovering of terrible remnants of industrial pollution in nations of the old Eastern bloc, to the rapidly expanding frontiers of residual sludges management and of nutrients and toxics control here in the West. Such immense environmental challenges can only be met through the marshalling of the talents of the best environmental engineers and scientists around the world and through the design and application of cost-effective solutions.

The International Association on Water Pollution Research and Control, the IAWPRC, is an organization capable of addressing these global challenges. The primary mission of the IAWPRC is to disseminate the results of both fundamental and practical research to the international professional community and to apply this information to the development of innovative and effective solutions to the problems of water pollution control. Technology transfer and reports of research and development are regularly published in the Proceedings of the IAWPRC's international biennial and specialized conferences, as well as in the IAWPRC journals: *Water Research, Water Science and Technology,* and *Water Quality International.*

Within the Association biennial conferences provide a unique forum for international researchers and practitioners to debate and discuss methods to address and resolve increasingly complex environmental problems. In keeping with the missions of the IAWPRC, the United States of America National Committee, USANC, organized eight Specialty Courses offered in conjunction with the 1992 IAWPRC Biennial Conference in Washington, D.C. Designed for the practicing engineer, the Specialty Courses covered critical topics in environmental quality management, water pollution control, wastewater treatment, toxicity reduction, and residuals management.

The eight volumes in the *Water Quality Management Library* represent the texts prepared for the Specialty Courses. Experts from the United States and many other countries contributed their know-how and experience to the preparation of these state-of-the-art texts. Collectively, the volumes in the library are a pertinent and timely compendium of water pollution control and water quality management. They form a unique reference source reflecting international expertise and practice in key aspects of modern water pollution science and technology.

W. Wesley Eckenfelder
Joseph F. Malina, Jr.
James W. Patterson

List of Contributors

PHILIP B. BEDIENT, PH.D., P.E., Rice University, Houston, Texas

RANDALL J. CHARBENEAU, PH.D., P.E., University of Texas at Austin, Austin, Texas

PAUL C. JOHNSON, PH.D., P.E., Shell Development Company, Houston, Texas

RAYMOND C. LOEHR, PH.D., P.E., University of Texas at Austin, Austin, Texas

HANADI S. RIFAI, PH.D., P.E., Rice University, Houston, Texas

CHAPTER 1

Introduction

THIS introductory chapter reviews the sources of soil and groundwater contamination and the potential remediation technologies. There are a number of remediation technologies that are commonly utilized in practice. These are the focus of the chapters that follow. In addition, there are a number of innovative (and sometimes expensive) technologies that show promise for special problem circumstances. Some of these are listed in this chapter for reference, though they will not be discussed in the later chapters.

SOURCES OF SUBSURFACE CONTAMINATION

The quality of subsurface waters may be impacted both by naturally occurring processes as well as by actions directly attributable to human activities. Pye and Kelley (1984) [1] note three general ways in which the chemical composition of groundwater may be changed. The first is due to natural processes. Leaching of natural geologic deposits can result in increased concentrations of chlorides, sulfates, nitrates, iron, and other inorganic chemicals. Evapotranspiration from shallow water tables can further concentrate salts in these aquifers. These problems are most common in the arid southwest and southcentral areas of the United States.

The second general source of groundwater contamination involves those activities associated with waste-disposal practices. In the United States, a comprehensive review of waste disposal effects on groundwater quality was undertaken as called for in passage of the Safe Drinking Water Act in 1974. This act directed the U.S. Environmental Protection Agency (EPA) to conduct a survey of waste disposal practices (including residential waste) that could endanger underground water that supplies, or could reasonably be expected to supply, any public water systems, and investigate means of control of such waste disposal. The resulting report to

1

Congress (U.S.EPA, 1977) [2] noted that waste disposal practices had affected the safety and availability of groundwater but that overall usefulness had not been diminished on a national basis. Importantly, this report also noted that instances of groundwater contamination are usually not discovered until after a drinking-water source had been affected. This points to the long-term problem of groundwater contamination: ground-water moves slowly so that it takes a long time for contaminants to appear at a potential receptor point, while the flip-side says that it will take a correspondingly long time to remediate the initial release of contaminants.

The waste disposal practices identified by the EPA survey include: (1) industrial wastewater that is contained in surface impoundments (lagoons, ponds, pits, and basins); (2) municipal and industrial solid refuse and sludge that are disposed of on land; (3) sewage wastes from homes and industries that are discharged to septic tanks and cesspools; (4) municipal sewage and storm-water runoff that are collected, treated, and discharged to the land; (5) municipal and industrial sludge that is land spread; (6) brine from petroleum exploration and development that is injected into the ground or stored in evaporation pits; (7) solid and liquid wastes from mining operations that are disposed of in tailing piles, lagoons, or discharged to land; (8) domestic, industrial, agricultural, and municipal wastewater that is disposed of in wells; and (9) animal feedlot waste that is disposed of on land and in lagoons. This list is fairly comprehensive and names most of the problem sources that we face today.

The third general source of groundwater contamination mentioned by Pye and Kelley (1984) [1] is also a direct result of human activities but is unrelated to waste disposal practices. This includes such things as accidental spills and leaks, agricultural activities, mining, salt-water intrusion, and others. Of these, perhaps the most significant problem is associated with leaking underground storage tanks. The majority of these tanks store petroleum products, principally heating and motor fuels, while others contain inorganic products such as acids and caustic products. In appreciation of the size of this problem, Plehn (1985) [3] notes that EPA estimates that the cost for cleanup of leaking underground tanks will approach $60 billion over the next thirty years.

There are a number of difficulties that arise when one attempts to determine the overall magnitude of the problem of groundwater contamination. Pye and Kelley (1984) [1] note that the definition of severity may be approached in several ways. A contamination incident may be considered severe if it results in groundwater concentrations that exceed standards set for drinking water, if the intended use is for water supply. If the ground-water was not intended for drinking, then the same incident might not be considered severe. One could also attempt to assesss severity by the number of persons affected by contamination, or by the percentage of an aquifer that is contaminated, or, on a nationwide basis, by the percentage of the available groundwater that is affected. Still other measures of severity arise if one includes the volume of contaminant released, its toxicity and persistence, and its mobility within the subsurface.

Regardless of what method is used, the amount of data necessary for assessing the severity of contamination in quantitative terms is not available. Experience has shown that it takes considerable effort and expense to establish the size and concentration distribution within a contaminant plume at a single site. For a national analysis, all one can do is extrapolate from the limited data we have obtained at the relatively few contaminated sites that have been properly characterized. Using available data, the U.S.EPA (1980) [4] has obtained an order-of-magnitude estimate of the extent of groundwater contamination from landfills and surface impoundments, which were considered to be the primary sources of subsurface contamination. Using information on whether the facilities are situated over useable groundwater, the length of time that the facilities have been operating, and the amount of groundwater in storage, they estimated that between 0.1% and 0.4% of the useable shallow aquifers are contaminated by industrial impoundments and landfill sites. The U.S.EPA (1980) [4] also investigated secondary sources such as septic tanks and petroleum exploration and mining. Extrapolating from the available data they concluded that such sources had contaminated about 1% and 0.1%, respectively, of the nation's useable shallow aquifers.

In another study, Lehr (1982) [5] assumed, for a worst-case scenario, (1) that 200,000 sources of contamination have operated for longer durations than in the EPA study and (2) that groundwater moves at faster rates than in the EPA study. For this worst-case scenario, he estimates that between 0.2% and 2% of the shallow groundwater has become contaminated, and he concludes that the fraction of polluted groundwater is small.

The conclusion from these and other investigations is that the total fraction of subsurface water that is contaminated is small. It is primarily the shallow aquifers that are susceptible to contamination, and there are vast quantities of good groundwater within deeper aquifers throughout the world. This does not mean to say that the problem of groundwater contamination is insignificant. Certainly, if your water supply well becomes contaminated, then it is a severe problem to you. In the United States, the primary goal of the Resource Conservation and Recovery Act (RCRA) is to prevent future releases of contamination from waste storage and disposal facilities, while the primary goal of the Comprehensive Environmental Response, Compensation, and Liability Act (CERCLA, or Superfund) is to clean up existing contaminated sites. Through these and

other pieces of legislation it is hoped that the problem of groundwater contamination has been placed under control and that the magnitude of the
problem will decrease over time.

CONTAMINATED SOIL AND GROUNDWATER REMEDIATION

Along with the widespread occurrence of problems associated with subsurface contamination has been the recognized need for development of
technologies for subsurface remediation. The choice of which method is
best for remediation of a given site, or whether remediation is even necessary, is difficult because the required data for assessing the environmental
consequences of groundwater contamination are not readily available and
the range of application of some of the remediation technologies is poorly
known. As outlined in Figure 1.1, the question of whether remediation is
necessary relies upon three pieces of information. First, one must know
whether there has been a release of contamination to the environment. For
problems discussed in this text it is generally assumed that a release has
occurred, though the amount and timing of the release may remain in
question. The second piece of information that is required concerns the
potential transport pathways from the point of release to the point of exposure to the general population. The material in this text is generally concerned with subsurface transport pathways, though for problems with contaminated soil, one may also look at volatile losses, resuspension of the
soil, and atmospheric transport. The third requirement for deciding
whether remediation is necessary at a site is a determination of whether
there is a potentially exposed population. If one examines the transport
pathways and finds that there is no potentially exposed population at the
various terminal points within the biosphere, then there is no risk to the
population and the need for remediation may be questioned.

The simplest and least expensive method of remediation is the no action
option. This may be called either natural or passive remediation, and it
relies upon natural processes, including biodegradation, volatilization,
and sorption, to remediate or contain the contamination. Charbeneau
(1984) [6] notes that with respect to in situ uranium leach mining, restoration based on natural processes has never been proposed by an operator
nor approved by any state or federal agency. The mechanisms of natural

Figure 1.1 Assessment of contamination impacts.

restoration are complex, and the important physicochemical properties are highly variable. This makes it difficult to provide evidence and assurance of improvement in groundwater quality by natural mechanisms. Despite these uncertainties, however, natural restoration may yet prove to be adequate to protect the environment.

The second easiest remediation option to apply, and that which was primarily used for contaminated soils through the mid-1980s, is to excavate the soil and transport it to an appropriate disposal site. The technology is simple, though it has fallen into disfavor because of limited landfill space and the costs and long-term liabilities involved. This technology results in moving a problem around, rather than in treating or containing the contamination onsite as encouraged in more recent legislation.

Subsurface remediation technologies that require some action on the part of the user may be broadly classified into two groups. The first group consists of those technologies that treat and/or contain the contaminants in place. The contaminant is not moved to another location for treatment. These may be called in situ technologies, with the two most common members being contaminated soil bioremediation and groundwater bioremediation. Biodegradation is the process by which the growth and activity of naturally occurring microorganisms is stimulated, and these organisms, through their metabolic processes, degrade the compounds of interest. Biodegradation of organic chemicals is a naturally occurring process within soil and groundwater. The goal of engineering a site is to enhance the rates of biodegradation by providing a more conducive environment for the organisms through management of oxygen, nutrients, water content, pH, and other factors. Contaminated soil biodegradation will be discussed in detail in Chapter 4, while groundwater biodegradation will be discussed in Chapter 6.

The second major group of subsurface remediation technologies consists of those that require displacement of the contaminant to another location for treatment or disposal. The most commonly applied members of this group include groundwater pump-and-treat systems, soil vapor extraction systems, and free product recovery systems.

By and large, the most common method of groundwater remediation has been application of pump-and-treat systems. Contaminated groundwater is recovered through production wells. At the ground surface the water is treated using air stripping, carbon adsorption, biological treatment, or other methods. If free product hydrocarbons are present, then an oil/water separator may be required. The treated water may be returned to the aquifer via injection wells or disposed of through surface drainage. Experience has shown that pump-and-treat systems can be very effective for removal of groundwater contaminants. However, this method is usually quite expensive and time consuming, and the total recovery time has often been significantly underestimated. Groundwater pump-and-treat systems are discussed in some detail in Chapter 5.

Soil vapor extraction (otherwise known as vacuum extraction, in situ volatilization, or soil venting) is the process by which volatile compounds are removed from the soil in the vadose zone through utilization of forced or drawn air currents. This technology is similar to the use of pump-and-treat systems in the saturated zone, except that the displacing fluid is air and that the effluent may or may not receive treatment (air stripping with activated carbon). Depending on the type of compounds and the nature of the host porous medium, soil venting can be a very effective, cost-efficient means of remediation. This remediation method is discussed in some detail in Chapter 7.

Leaking storage tanks and pipelines can result in subsurface contamination with hydrocarbons that remain immiscible with water. These are the non-aqueous phase liquids (NAPLs) or organic immiscible liquids (OILs), and they may be further subdivided into those that float on the water table (light NAPLs or LNAPLs) such as gasoline or fuel oil, and those that sink through the saturated zone (dense NAPLs or DNAPLs) such as chlorinated solvents. The recovery of free product hydrocarbon floating on the water table is similar in concept to the pump-and-treat systems, but there are sufficient differences so that this problem is treated separately in Chapter 8. Unfortunately, there are no good methods for recovery of DNAPLs.

There are a number of other remediation technologies, many of which were reviewed as part of a cooperative effort between the Electric Power Research Institute (EPRI) and the Utility Solid Waste Activities Group (USWAG). The EPRI (1988) [7] study provides a description and evaluation of available technologies for remediating soil and groundwater that contain petroleum products released from underground storage tank leaks, other discharges, or spills. These include the in situ technologies of leaching, vitrification, and isolation/containment and the non-in situ technologies of land treatment, thermal treatment, asphalt incorporation, solidification and/or stabilization, chemical extraction, and excavation.

In situ leaching is the process by which soils are washed or flushed with water and are usually mixed with surface active chemicals in an effort to leach the compounds present within the soil matrix downward to groundwater. The contaminants are then captured along with groundwater through a collection system located either beneath or downgradient of the leaching site and are transported to a surface system for treatment and disposal. In situ leaching is not commonly practiced and is not discussed further.

In situ vitrification is a thermal treatment process by which soils are vitrified through utilization of electricity. Heat generated by the passage of electricity through the contaminated soil converts the soil into a chemically inert and stable glass and crystalline product. During the vitrification process, the major portion of the compounds initially present in the soils are volatilized with the remainder being worked in place in the hardened soil. In situ vitrification was originally developed at Pacific Northwest Laboratories for the purpose of stabilizing high-level radioactive wastes in place. The technology is relatively new and utilizes large amounts of energy. The relative costs are high.

As the name implies, in situ isolation/containment refers to the separation of the contaminated region from the rest of the environment using such devices as caps, grout curtains, and cut-off and slurry walls. The goal of isolation is to immobilize the contaminant within a given region and prevent its offsite migration. Experience has shown that isolation and containment systems can work adequately, but that destruction of the contaminants is not accomplished.

The non-in situ technologies involve the removal of the contaminated soil and groundwater from the immediate site of contamination for treatment. One such technology is land treatment, which is a managed technology that involves controlled application of a waste on the soil surface or incorporation of the waste into the upper soil zone. The site is engineered to promote processes that result in waste degradation and reduce mobility. Land treatment is also used for contaminated soil bioremediation, as discussed in Chapter 4.

During thermal treatment or incineration, the contaminated soil is removed from the site and exposed to excessive heat, resulting in the loss of organic chemicals through volatilization or other thermal destructive reactions. Both low-temperature and high-temperature incinerators are available. Incineration is effective for removal of hydrocarbon compounds in soils. However, the associated costs may be significant.

Chemical extraction or soil washing is the process by which excavated soils are washed using water containing solvents or surfactants to remove the compounds of concern. The technology is similar to in situ leaching except that the soil is not treated in place, allowing wash mixtures to be used that would not be environmentally safe if applied in situ. Soil washing has shown promise for removal of heavy metals and radionuclides, as well as heavier organic compounds.

Finally we mention two additional technologies that are discussed in EPRI (1988) [7]. These are incorporation of the contaminated soil in asphalt as a substitute for aggregate, and solidification/stabilization of the contaminated soil through incorporation of additives to encapsulate the compounds of concern. With regard to asphalt incorporation, the volatile compounds are driven off during heating of the asphalt mixture. The remaining compounds are rendered less mobile after the asphalt has set. Neither of these technologies has found wide application because of their expense and technical uncertainties.

Table 1.1 provides a summary list of the important remediation tech-

TABLE 1.1. Remediation Technologies for Contaminated Soils and Groundwater.

In situ technologies	Non-in situ technologies
contaminated soil bioremediation	pump and treat
groundwater bioremediation	vapor extraction
leaching and chemical reaction	free product recovery
vitrification	land treatment
isolation/containment	incineration
	soil washing
	asphalt incorporation
	solidification/stabilization
	excavation

nologies for contaminated soils and groundwater. The most commonly applied methods include bioremediation of contaminated soil and groundwater (including land treatment), pump-and-treat systems, vapor extraction, free product recovery, incineration, and excavation. Except for incineration and excavation, these are discussed in later chapters. However, Chapters 2 and 3 first discuss the important processes that control movement of subsurface water and the transport of contaminants.

REFERENCES

1 Pye, V. I. and J. Kelley. 1984. "The Extent of Groundwater Contamination in the United States," in *Groundwater Contamination*, National Academy Press, Washington, D.C.

2 U.S. Environmental Protection Agency. 1977. "Waste Disposal Practices and Their Environmental Effects on Ground Water," Report to Congress, Washington, D.C., 512 pp.

3 Plehn, S. W. 1985. *Underground Tankage, the Liabilities of Leaks*, Petroleum Marketing Education Foundation, Alexandria, VA.

4 U.S. Environmental Protection Agency. 1980. *Planning Workshops to Develop Recommendations for a Ground Water Protection Strategy, Appendixes*, Washington, D.C., 171 pp.

5 Lehr, J. H. 1982. "How Much Ground Water Have We Really Polluted?" (editorial), *Ground Water Monitoring Rev.* (Winter), 4–5.

6 Charbeneau, R. J. 1984. "Groundwater Restoration with In Situ Uranium Leach Mining," in *Groundwater Contamination*, National Academy Press, Washington, D.C.

7 Electric Power Research Institute. 1988. *Remedial Technologies for Leaking Underground Storage Tanks*, Chelsea, MI, Lewis Publishers.

CHAPTER 2

Darcy's Law and Subsurface Flow

THE potential fate and transport of subsurface contaminants, as well as the operation of many remediation systems, depend to a great extent on the flow of water and on the mobility and partitioning characteristics of contaminants within the multiphase subsurface environment. This chapter reviews the principles of subsurface fluid movement. While the greatest interest focuses on the flow of water in saturated porous media, the flow of water and air in the vadose zone is also discussed. The transport of contaminants is considered in Chapter 3.

DARCY'S LAW

The basis for calculation of fluid flow in porous media is called Darcy's law. This law expresses a balance between the pressure gradients and gravity that drive the flow and the viscous resistance to fluid motion. Because of the small pore dimensions and the small fluid velocities, subsurface flow is usually laminar rather than turbulent. This means that the viscous forces are proportional to the velocity to the first power, and Darcy's law expresses a linear relationship between the energy gradient which causes the flow and the flow velocity.

Using an experimental apparatus similar to that shown in Figure 2.1, Darcy arrived at the following empirical law relating the total discharge, Q, across the filter bed to its area, A, water level change across the filter, $z_1 - z_2$, and filter thickness, L.

$$Q = KA \frac{z_1 - z_2}{L} \qquad (2.1)$$

The constant K appearing in Equation (2.1) is called the hydraulic conductivity, though in older literature the term coefficient of permeability

9

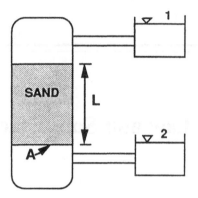

Figure 2.1 Experimental apparatus similar to that used by Darcy.

was often used. A look at the dimensions of all the terms shows that K has the same units as velocity. Equation (2.1) may be written in a more general and useful form if one recognizes that

$$\frac{z_1 - z_2}{L} = I$$

is the hydraulic gradient across the sand. Further, the specific discharge, or Darcy velocity, is defined as the total discharge from the entire cross-sectional area:

$$U = \frac{Q}{A}$$

With these changes Darcy's result may be written

$$U = KI \tag{2.2}$$

Typical values and ranges of the hydraulic conductivity are shown in Table 2.1. The important point to notice from this table is that for natural porous media, K varies by over 12 orders of magnitude. There are no other physical parameters that show this great range of variation. Variations of 3 orders of magnitude are common even for a single formation.

HYDRAULIC HEAD AND GRADIENT

In fluid mechanics the mechanical energy content of a fluid is specified by its hydraulic head, h, where

$$h = \frac{v^2}{2g} + \frac{p}{\gamma} + z \tag{2.3}$$

The hydraulic head, or just the head, is the mechanical energy per unit weight and has the dimensions of length. Equation (2.3) shows that the head is the sum of the velocity head, $v^2/2g$, which is the kinetic energy per unit weight, the pressure head, p/γ, which is a measure of the ability of the fluid to do work, and the elevation head, z, which is the potential energy per unit weight. For problems of subsurface flow the velocity head is neglible so

$$h \cong \frac{p}{\gamma} + z \qquad (2.4)$$

The right-hand side of Equation (2.4) is the piezometric head, which is the level to which water will rise in a piezometer. Since the velocity head is small for groundwater flow problems, the hydraulic head and piezometric head are essentially the same.

PIEZOMETERS

A piezometer is the basic field device used for measurement of the groundwater hydraulic head. As shown in Figure 2.2, it consists of a tube or pipe in which the water level can be measured. Water will enter or leave the screened section until the energy in the piezometer comes into equilibrium with that within the porous medium.

If measurements from a number of piezometers are available, then one can plot and contour the "piezometric surface." Piezometers that are screened at the same level can be used to determine the horizontal direction of flow as shown in Figure 2.3, while a set that is screened at different

TABLE 2.1. Typical Value of Hydraulic Conductivity (after Marsily [1]).

Medium	K (cm/s)
Unconsolidated material	
Coarse gravels	10^1–10^0
Sands and gravels	10^0–10^{-3}
Fine sands, silts, and loess	10^{-3}–10^{-7}
Clay, shale, glacial till	10^{-7}–10^{-11}
Unfractured rocks	
Dolomitic limestones	10^{-1}–10^{-3}
Weathered chalk	10^{-1}–10^{-3}
Unweathered chalk	10^{-4}–10^{-7}
Limestone	10^{-3}–10^{-7}
Sandstone	10^{-2}–10^{-8}
Granite, gneiss, compact basalt	10^{-7}–10^{-11}

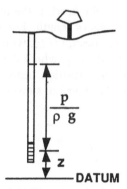

Figure 2.2 A field piezometer for measurement of the hydraulic head.

levels but at the same horizontal location may be used to determine vertical gradients. If a long screened interval is used, then the water level in the piezometer gives an average value of the head. For site characterization work it is important to screen only a narrow interval.

HOMOGENEOUS AND ISOTROPIC MEDIA

The hydraulic conductivity is perhaps the most important property of a porous medium. A value of K can be assigned to every point within a formation, and one can then refer to the hydraulic conductivity field. If the value of K is the same at every point, then the field is said to be *homogeneous*; otherwise, it is a *heterogeneous* field. Under the most general conditions, the hydraulic conductivity also has directional characteristics, with the conductivity being larger in some directions than in others. In this case the field is *anisotropic*. If the magnitude of K is independent of direction, then the field is *isotropic*. Thus, in the most general case, the

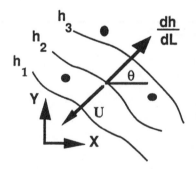

Figure 2.3 Piezometric surface with $h_1 < h_2 < h_3$.

hydraulic conductivity field is both heterogeneous and anisotropic. For engineering calculations at a particular site one often assumes that the hydraulic conductivity is independent of direction in the horizontal plane. Stratification that is typical of most geologic deposits causes the hydraulic conductivity to be less in the vertical direction than in the horizontal direction. This simple anisotropy is an adequate description of field sites for most applications.

FACTORS THAT INFLUENCE K

Detailed studies of porous media flow show that the hydraulic conductivity of the medium is a function of both fluid and medium properties, expressed by

$$K = \frac{k\varrho g}{\mu} \tag{2.5}$$

In Equation (2.5), k is the intrinsic permeability of the medium. The intrinsic permeability is assumed to be a function only of the porous matrix. Its value depends on the porosity, pore size distribution, soil texture and structure, and possibly other factors. The intrinsic permeability is not a function of the invading fluid, so long as this fluid does not change the soil structure. The fluid properties that appear in Equation (2.5) are the water density ϱ and the dynamic viscosity μ. The parameter g is the gravitational constant. The density ϱ depends only slightly on temperature. However, density variations due to solute concentrations can be significant, especially when dealing with problems of salinity intrusion or disposal. On the other hand, the water viscosity is sensitive to temperature variations and this can be significant in climates that show a wide range of temperatures over the years.

When the density of water varies from point to point, because of differences in salinity for example, then the entire form of Darcy's law must be modified and one cannot work with the hydraulic conductivity of the medium or the hydraulic gradient. The most general form of Darcy's law is

$$U = -\frac{k}{\mu} (\nabla p + \varrho g \mathbf{k}) \tag{2.6}$$

where ∇p is the pressure gradient and $\mathbf{k}$ is the vertical upward unit vector. This form must be used when dealing with problems of salt-water intrusion, for example. It is also the standard form used by petroleum engineers, who must deal with multiple fluids occupying the pore space.

Another factor that has an important influence on the hydraulic conductivity of a porous medium is the cation composition for clayey soils. Many clay soils can exhibit either a flocculated or dispersed structure, depending on the type of cations present and the ionic strength of the fluid. The structure that exists is associated with the double layer surrounding the negatively charged clay particles. It takes fewer higher valence cations in the double layer to balance the negative charge of the clays, and the resulting double layer is thin. Such soils tend to flocculate and form aggregates, leading to good hydraulic characteristics. On the other hand, low valence cations, such as Na, result in expanded double layers and dispersed clays. Such soils show poor structure with low permeabilities and they do not drain well. This picture is complicated to some extent in that a soil solution with a high ionic strength can compress the double layer and lead to good structure, even if the dominant cation is Na. Such is the case for a soil invaded by seawater.

LIMITATIONS TO THE VALIDITY OF DARCY'S LAW

Meinzer (1942) [2] notes that there does not appear to be a lower limit to the range of applicability of Darcy's law for aquifer materials. With hydraulic gradients of as small as a few inches per mile, the linear relationship remains valid. For clay materials, on the other hand, the situation is not so clear. Because of the large charged surface area of clays, there is some suggestion that water does not act as a Newtonian fluid, at least for small gradients, and that there is a threshold gradient that must be exceeded before flow occurs. For larger hydraulic gradients and velocities there is ample evidence that the flow does depart from the linear relationships of Darcy's law. At these large flow rates, inertial effects become important, and the flow characteristics approach those of turbulent flow. Experiments have shown that these effects do not become important until the Reynolds number, N_R, reaches a value of about 1 to 10 where

$$N_R = \frac{U \varrho d}{\mu}$$

and where d is the mean or effective grain size. Laminar flow conditions are met under most field applications and it is generally assumed that Darcy's law applies throughout.

LABORATORY MEASUREMENT

Laboratory measurement of permeabilities involves the use of standing head and falling head permeameters. A standing head device is similar to

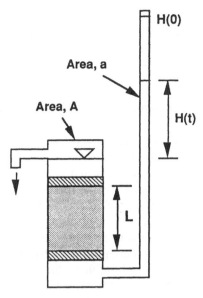

Figure 2.4 Falling head permeameter used to measure hydraulic conductivities of low permeability samples.

that used originally by Darcy and shown in Figure 2.1. A constant head difference is maintained across the soil sample and the flow rate is measured (alternatively, the flow rate is fixed and the head difference is measured). Such a device is most often used for samples with appreciable permeability so that the flow rate is easily measured. Equation (2.1) may be rewritten to give

$$K = \frac{QL}{A(h_1 - h_2)} \qquad (2.7)$$

where h_1 and h_2 are the upstream and downstream hydraulic heads, respectively, and L is the length of the packed soil column.

A falling head permeameter, such as that shown in Figure 2.4, is often used for permeability measurement in low permeability soils where the discharge rate is small. A constant effluent head is maintained and the water level in the small standpipe is measured as a function of time. Measurement of permeability follows from application of continuity principles, which lead to

$$K = \frac{La}{A(t - t_0)} \ln\left(\frac{H_0}{H}\right) \qquad (2.8)$$

Thus the permeability may be estimated either from individual measurements of the water level in the standpipe or from the slope of a plot of the logarithm of H versus time. Field methods for measurement of the hydraulic conductivity are discussed later.

CONTINUITY EQUATION FOR GROUNDWATER HYDRAULICS

The continuity equation is a mathematical statement of the physical law of conservation of mass. Together, Darcy's law and the continuity equation, along with appropriate boundary and initial conditions, provide the mathematical framework to solve for the head and velocity throughout a domain as a function of location and time. The basic continuity principle from fluid mechanics states that for an arbitrary control volume, the rate of mass accumulation within the volume plus the net mass flux out of the volume must equal the rate of mass generation within the volume (Bird et al., 1960) [3]. For most practical problems the changes in the density of water are small, and to the first order, the density of water may be taken as constant. This means that instead of stating a law of conservation of mass, the continuity equation as it is normally written states a condition of conservation of volume of water. For an arbitrary control volume, the continuity equation states

$$
\begin{array}{ccc}
\text{rate of volume} & \text{net} & \text{volume} \\
\text{storage} + \text{volume} = \text{source} \\
\text{increase} & \text{flux out} & \text{strength}
\end{array}
\qquad (2.9)
$$

For a confined aquifer the continuity equation takes the form

$$
S \frac{\partial h}{\partial t} = \frac{\partial}{\partial x}\left(T \frac{\partial h}{\partial x} \right) + \frac{\partial}{\partial y}\left(T \frac{\partial h}{\partial y} \right)
\qquad (2.10)
$$

where S is the coefficient of storage and T is the aquifer transmissivity. The mechanism of storage release in a confined aquifer is associated with the compressibility of the porous matrix and of water. If a well is turned on and produces water from the aquifer, then the pressure at a given location will decrease. Since the total weight of soil and water above the location remains the same, the fact that the pressure is decreasing means that this weight must be supported by an increase in the effective stress within the porous matrix. An increase in effective stress results in a compression (consolidation) of the aquifer. However, since the individual soil grains have a very low compressibility, the resulting bulk volume change occurs primarily from a reduction in porosity of the aquifer. At the same time this is happening, the decrease in pressure results in a slight decrease in water density (expansion of the water). Both of these effects make it look like there is more water present at the given location. One says that water has been released from "storage" associated with the decrease in pressure (head). This is the meaning of the first term in Equation (2.9), and one may define the storage coefficient as the volume of water released from or added to storage per unit change in head per unit area (in the plan view):

$$S = \frac{1}{A} \frac{dV_w}{dh} \tag{2.11}$$

The transmissivity, T, is a parameter that represents the ability of the aquifer as a whole to transmit water under a given horizontal gradient in head. It is equal to the integral of the hydraulic conductivity over the thickness of the aquifer, or roughly,

$$T = Kb \tag{2.12}$$

where b is the aquifer thickness. Thus the discharge in the x-direction, per unit width of the aquifer, is given by

$$U_x b = -T \frac{\partial h}{\partial x} \tag{2.13}$$

and similarly for the y-direction.

FLOW TO A PRODUCTION WELL

The continuity equation provides the basis for development of numerical models for simulation of groundwater flow. Under somewhat restrictive conditions, it may also be solved analytically. Perhaps the most important analytical solution was presented by Theis (1935) [4]. For the situation shown in Figure 2.5, the following assumptions are made: (1) The aquifer is homogeneous, isotropic, and large (infinite). (2) The original piezometric surface is horizontal. (3) The well is pumped at a constant discharge. (4) The well is fully penetrating and the flow is horizontal. (5) The well diameter is infinitesimal so that storage within the well can be neglected. (6) The aquifer storage release is instantaneous. With these assumptions the solution to the continuity equation in a confined aquifer is given by

$$h_0 - h(r, t) = \frac{Q}{4\pi T} \int_u^\infty \frac{1}{w} e^{-w} dw \tag{2.14}$$

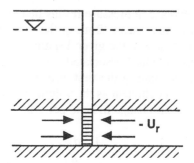

Figure 2.5 Configuration for flow to an ideal well in a confined aquifer.

where w is the dummy variable of integration and

$$u = \frac{r^2 S}{4Tt} \tag{2.15}$$

The integral in Equation (2.14) is known in mathematical physics as the exponential integral. In groundwater hydraulics it is called the Theis Well Function and designated

$$W(u) = \int_u^\infty \frac{1}{w} e^{-w} dw \tag{2.16}$$

The Theis well function is tabulated in many references (see for example, Freeze and Cherry, 1979 [5]). The difference $h_o - h(r, t) = s(r, t)$ is called the *drawdown*, and Equation (2.14) may be written

$$s(r, t) = \frac{Q}{4\pi T} W(u) \tag{2.17}$$

This is the *Theis equation* and describes the development of a "draw-down cone" as a function of radius and time.

For small r or large t, u is small, and Cooper and Jacob (1946) [6] noted that an approximation to the Theis well function could be written as

$$s(r, t) = \frac{Q}{4\pi T} \ln \left(\frac{2.25Tt}{r^2 S} \right) \tag{2.18}$$

This is called the *Jacob equation*. This approximation is useful so long as $u < 0.01$. In order to see how to use the Jacob equation, consider Example 2.1.

Example 2.1

An aquifer is pumped at a rate of $Q = 1500$ m³/d for 1 year. Estimate the drawdown at the well of radius $r_w = 0.2$ m and at a distance of 1 km if the aquifer has a transmissivity of 600 m²/d and a storage coefficient of $S = 0.0004$.

For the well one finds

$$u = \frac{0.2^2 \times 0.0004}{4 \times 600 \times 365} = 1.8 \times 10^{-11}$$

so that the Jacob approximation is valid. This approximation then gives

$$s(0.2, 365) = \frac{1500}{4\pi 600} \ln\left(\frac{0.561}{1.8 \times 10^{-11}}\right) = 4.81 \text{ m}$$

Similarly, at 1 km the approximation is also valid and one finds

$$s(1000, 365) = \frac{1500}{4\pi 600} \ln\left(\frac{0.561}{4.6 \times 10^{-4}}\right) = 1.41 \text{ m}$$

The radius of influence might be described as the distance from a pumping well to where the drawdown is negligible. Here, the term "negligible" is problem dependent, and so the radius of influence is not a readily fixed quantity. A useful representation is found through Jacob's approximation. For $s = 0$, the argument of the logarithmic function must equal unity. Using the radius that satisfies this equation as the radius of influence, R, one has

$$R = 1.5\sqrt{\frac{Tt}{S}} \tag{2.19}$$

The radius of influence of a well increases with the square root of time and is larger in formations with greater transmissivities and smaller storage coefficients. It is important to note that the radius of influence is independent of the discharge. At first glance this does not appear to be correct. However, this radius is a measure of the distance of propagation of small energy disturbances, and this small disturbance limit is independent of the discharge. The Jacob equation shows that within the radius of influence, the drawdown is directly proportional to Q.

Substitution of Equation (2.19) into the Jacob equation leads to

$$s = \frac{Q}{4\pi T} \ln\left(\frac{R^2}{r^2}\right) = \frac{Q}{2\pi T} \ln\left(\frac{R}{r}\right) \qquad (2.20)$$

This equation is very important in its own right. It is called the *Thiem* equation and it may be developed as the steady-state solution for flow to a well in a confined aquifer. A slightly more general form of Equation (2.20) may be found by considering the drawdowns at two different radii. This gives

$$s_2 - s_1 = \frac{Q}{2\pi T} \ln\left(\frac{r_1}{r_2}\right) \qquad (2.21)$$

Equation (2.21) may be used to estimate the transmissivity of an aquifer by measuring the drawdown in two observation wells at radii r_1 and r_2. Thiem used this idea to calculate the transmissivity as

$$T = \frac{Q}{2\pi(s_2 - s_1)} \ln\left(\frac{r_1}{r_2}\right) \qquad (2.22)$$

Equation (2.22) is actually valid both for steady and unsteady flow conditions so long as the condition $u < 0.01$ is valid for unsteady flow at both observation wells and the drawdowns are measured at the same time.

AQUIFER TESTS

The determination of aquifer characteristics through aquifer tests has become a standard step in the evaluation of groundwater resource potential and in the design of contaminant recovery systems. In practice, there is much art to successful aquifer testing and data interpretation, and an excellent reference is Kruseman and de Ridder, 1970 [7]. In addition, many case histories are presented by Walton, 1970 [8]. The basic idea in an aquifer test is to pump one well and measure its discharge and the drawdown in nearby observation wells. Figure 2.6 shows a possible well configuration and the type of data one obtains.

Estimation of the aquifer's transmissivity, storage coefficient, and other characteristics from an aquifer test involves direct application of the equations listed in this section. The response of an aquifer to pumping will depend on its characteristic parameters, and the method of aquifer testing involves either graphical or computer estimation of the parameters values from the observed drawdown and well production rate records. While computer methods of parameter estimation are preferred for the final analysis of data because they are analyst independent, the graphical methods

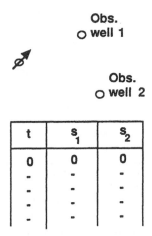

Figure 2.6 Aquifer test data.

are still important for practical and educational purposes. In particular, the semilogarithmic method should be used as a standard tool in field investigations to assess the performance of the aquifer test and to choose its completion time and recovery time. For confined aquifers the two graphical methods include the log-log plots of drawdown versus time for the *Theis method* and the semilog plots for the *Jacob method*. Only the Jacob method is discussed below.

Jacob's equation may be written

$$s = \frac{Q}{4\pi T} \ln t + \frac{Q}{4\pi T} \ln \left(\frac{2.25\, T}{r^2 S} \right)$$

which has the form of a straight line ($s = A \ln t + B$) on semilog paper with

$$\text{slope} = \frac{Q}{4\pi T}$$

Thus a plot of drawdown versus time on semilog paper should appear as a straight line, as shown in Figure 2.7. The slope of the line provides the transmissivity through

$$T = \frac{Q}{4\pi} \frac{\ln \dfrac{t_2}{t_1}}{s_2 - s_1} \tag{2.23}$$

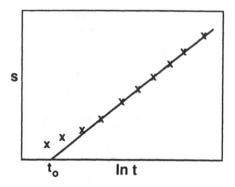

Figure 2.7 Semilog plot of drawdown versus time.

With the transmissivity provided by Equation (2.23), the $s = 0$ intercept may be used to estimate the storage coefficient. For the intercept, one finds

$$\ln t_0 = -\frac{B}{A} = -\ln\left(\frac{2.25\ T}{r^2 S}\right)$$

or

$$\ln\left(\frac{2.25\ T t_0}{r^2 S}\right) = 0$$

For the logarithmic function to equal zero, its argument must equal unity. Thus

$$S = \frac{2.25\ T t_0}{r^2} \tag{2.24}$$

Determination of T and S from a semilog plot with Equations (2.23) and (2.24) is called Jacob's method. One should note that the early-time data is not expected to fall on the straight line because it may not satisfy the assumptions on which the method is based, namely $u < 0.01$. Example 2.2 shows application of the Jacob method for aquifer test analysis.

Example 2.2

A well is pumped at a rate of 500 gallons per minutes for 500 minutes, and then the water level recovery is measured for an additional 500 minutes. The water level changes in an observation well located 40 ft from the production well are recorded and are shown in Figure 2.8. The aquifer's transmissivity and storage coefficient are to be estimated from this data.

The same data are plotted on a semilog graph in Figure 2.9. A straight

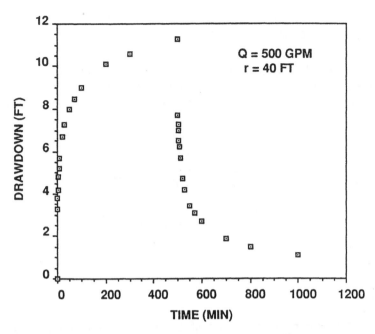

Figure 2.8 Drawdown and recovery data from an aquifer test.

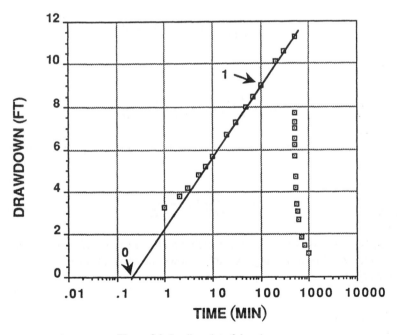

Figure 2.9 Semilog plot of drawdown.

23

line is drawn through the data points and two points are read off this line. These two data points are: $t_0 = 0.2$ min $= 0.00014$ days, $s_0 = 0$ ft and $t_1 = 100$ min, $s_1 = 9$ ft. With these two points, Equations (2.23) and (2.24) give

$$T = \frac{96260}{4\pi} \frac{\ln \dfrac{100}{0.2}}{9 - 0} = 5290 \text{ ft}^2/\text{d}$$

$$S = \frac{2.25 \, Tt_0}{r^2} = 0.0010$$

RECOVERY TESTS

Once the pumping well is shut off, the water levels start to recover. The drawdown during recovery is found from the principle of superposition. If the time of pumping is Δt, then

$$s = \frac{Q}{4\pi T} [W(u_1) - W(u_2)]$$

where

$$u_1 = \frac{r^2 S}{4Tt}, \, u_2 = \frac{r^2 S}{4T(t - \Delta t)}$$

A short period after the start of recovery, the Jacob equation becomes valid and

$$s = \frac{Q}{4\pi T} \ln \frac{t}{t - \Delta t} \tag{2.25}$$

A plot of s versus $t/(t - \Delta t)$ is called a *Horner plot* and can be used to calculate T. The following example shows an application of analysis of recovery data.

Example 2.3

The recovery data from the previous example are presented as a Horner plot in Figure 2.10. From the line drawn through the data points one may

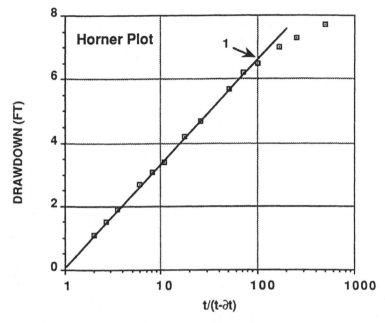

Figure 2.10 Horner plot of the recovery data.

read off the point $s = 6.6$ ft at $t/(t - \Delta t) = 100$. Using this value in Equation (2.25) one finds

$$T = \frac{Q}{4\pi}\frac{\ln\left(\dfrac{t}{t - \Delta t}\right)}{s} = \frac{96260}{4\pi}\frac{\ln 100}{6.6} = 5,340 \text{ ft}^2/\text{d}$$

Note that S cannot be calculated from the recovery data.

MULTIPLE OBSERVATION WELLS

Often one will have drawdown data from more than one observation well. Then it is possible to carry through the aquifer test analysis of each well separately. If the drawdown data are plotted with s versus t then the curves for the different wells are parallel. However, a look at both the Theis equation and the Jacob equation shows that t and r appear only in the combination t/r^2. Thus if the drawdown data are plotted as s versus t/r^2, then all of the data curves should fall upon each other. This is shown schematically in Figure 2.11. If this does not happen, then one obtains further information as to heterogeneities in the aquifer.

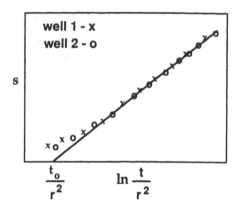

Figure 2.11 Semilog plot of s versus t/r^2 showing data sets as a single curve.

SLUG TESTS

On many occasions it is not convenient to run an aquifer test that requires continuous pumpage to evaluate aquifer parameters. For example, in low permeability units the well might not yield sufficient discharge for the pump to continue to operate. When characterizing a hazardous waste site, a conventional aquifer test would require treatment and proper disposal of all of the water produced. In these types of cases it is much more convenient to develop alternative testing methods.

Slug tests, borehole tests, rate-of-rise techniques, etc., provide an alternative testing method – though with potentially significant limitations. The basic idea is that a certain volume or "slug" of water is suddenly removed from a well, and the rate of rise of the water level in the well is measured. From this rate of recovery, one can estimate the permeability character of the formation, and possibly its storage characteristics also, though generally with less precision. Essentially instantaneous lowering of the water level in a well can be achieved by quickly removing the water with a bailer. Another method of lowering the water level which may be more appropriate at hazardous waste sites where the water may contain various contaminants is to submerge an object in the water, let the water level return to equilibrium, and then quickly remove the object. If the aquifer is very permeable, the water level in the well may rise very rapidly. If this is the case then the level may be measured by using sensitive pressure transducers and fast-response recording equipment. This section looks at two of the most widely used slug tests in practice. For further discussion and references one might consult Bouwer (1978) [9] and Bouwer and Jackson (1974) [10].

One of the earlier studies of borehole methods which is still widely used is that of Hvorslev (1951) [11]. Hvorslev was primarily concerned with use of water level measurements in boreholes to obtain pore-water pressures for design of foundations and earth structures. His investigations of the time-lag before useful measurements can be made have led to a method for estimation of permeability within the vicinity of the borehole, since the permeability is the primary control on the water flow rate near the observation point and the rate of pressure equilibration.

The general configuration for application of Hvorslev's analysis is shown in Figure 2.12. For $t < 0$ the water level in the borehole is at an elevation H above the datum. At time $t = 0$ the water level is dropped to an elevation H_0. The water level will ultimately recover again to the elevation H, while the water level at any time t is $h(t)$.

A simple application of the continuity equation shows that the relation between the water level within the borehole and the inflow rate, Q, is

$$Q = A\frac{dh}{dt}$$

where A is the cross-sectional area of the borehole. Application of Darcy's law gives

$$Q = FK(H - h)$$

where F is a factor that depends on the shape and dimensions of the intake or well point, K is the hydraulic conductivity in the vicinity of the intake, and $(H - h)$ is the potential difference between fluid outside of the borehole away from the intake and that within the borehole. Combining these equations gives

$$\frac{dh}{H - h} = \frac{FKdt}{A}$$

Hvorslev defines the *basic time lag* as the quantity

$$\tau = \frac{A}{FK} \tag{2.26}$$

In order to give the basic time lag a physical interpretation, note that it may be written

$$\tau = \frac{A(H - H_0)}{FK(H - H_0)} = \frac{V}{Q_0}$$

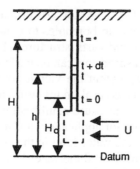

Figure 2.12 Configuration for Hvorslev method.

where V is the displaced volume of water and Q_0 is the initial inflow rate. Thus, τ is the time that would be required to re-establish equilibrium if the inflow rate remained constant at its initial value. In terms of τ Hvorslev finds

$$\frac{H - h}{H - H_0} = e^{-t/\tau} \tag{2.27}$$

Equation (2.27) shows that at $t = \tau$,

$$\frac{H - h}{H - H_0} = e^{-1} = 0.37$$

Thus, τ may be estimated by plotting $\ln\left[(H - h)/(H - H_0)\right]$ versus t and reading off the value of t where $(H - h)/(H - H_0)$ equals 0.37 (or $\ln\left[(H - h)/(H - H_0)\right] = -1$). With the value of τ known or estimated, then Equation (2.27) gives the hydraulic conductivity as

$$K = \frac{A}{F\tau} \tag{2.28}$$

This is the basis of the Hvorslev method for estimating permeability. It is very similar to the falling head permeameter mentioned earlier in this chapter.

Hvorslev presents the shape factor F for a number of configurations. For a spherical cavity at the end of a borehole he gives

$$F = 4\pi R \tag{2.29}$$

where the configuration is shown in Figure 2.13(a). For a hole extended in a uniform soil he presents

$$F = \frac{2\pi L}{\ln\left[\frac{L}{D} + \sqrt{1 + \left(\frac{L}{D}\right)^2}\right]} \tag{2.30}$$

and again the configuration is shown in Figure 2.13(b). For $L > 8D$, Equation (2.30) may be approximated by

$$F = \frac{2\pi L}{\ln\left(\frac{L}{r}\right)} \tag{2.31}$$

where $r = D/2$ is the hole radius. A number of other configurations are also presented by Hvorslev [11].

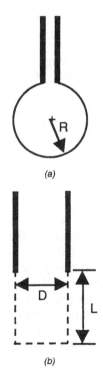

(a)

(b)

Figure 2.13 Configuration for shape factors: (a) spherical cavity and (b) hole extended in uniform soil (from Hvorslev, 1951 [11]).

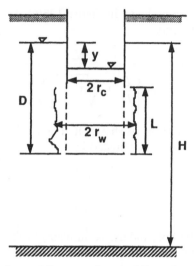

Figure 2.14 Geometry of a partially penetrating, partially perforated well in an unconfined aquifer with a gravel pack or developed zone around the perforated section.

A second method for analyzing slug test data in unconfined aquifers was presented by Bouwer and Rice (1976) [12]. The geometry for a well in an unconfined aquifer is shown in Figure 2.14. For the slug test the water level in the well is suddenly lowered, and the rate of rise of the water level is measured. The flow rate into the well may be estimated from the Thiem equation as

$$Q = \frac{2\pi KLy}{\ln (R_e/r_w)} \qquad (2.32)$$

where L is the height of the portion of the well through with water enters (height of screen or perforated zone), y is the vertical distance between the water level in the well and the equilibrium water table elevation, R_e is the effective radius over which y is dissipated (effective radius of influence), and r_w is the horizontal distance from the well center to the original aquifer (well radius plus thickness of the gravel envelope or developed zone). This equation assumes that the drawdown around the well is negligible, as are the head losses as water enters the well (well losses).

The rate of rise of the water level in the well after suddenly removing a slug of water is related to the inflow rate by continuity as

$$\frac{dy}{dt} = -\frac{Q}{\pi r_c^2} \qquad (2.33)$$

where πr_c^2 is the cross-sectional area of the well where the water level is rising. The term r_c is the inside radius of the casing if the water level is above the perforated or screened portion of the well. If the water level is rising in the perforated section of the well, and if the hydraulic conductivity of the gravel envelope or developed zone is much larger than that of the aquifer, then allowance must be made for the porosity outside the well casing. Bouwer and Rice (1976) [12] present an example showing that if the radius of the perforated casing is 20 cm and the casing is surrounded by a 10-cm permeable gravel envelope with a porosity of 0.30, then r_c should be calculated as

$$\sqrt{20^2 + 0.30 \times (30^2 - 20^2)} = 23.5 \text{ cm}$$

The value of r_w for this well section is 30 cm.

Equations (2.32) and (2.33) may be combined and integrated with the initial condition that $y = y_0$ at $t = 0$. This gives for the permeability

$$K = \frac{r_c^2 \ln (R_e / r_w)}{2L} \frac{1}{t} \ln \left(\frac{y_0}{y_t} \right) \tag{2.34}$$

Equation (2.34) is the solution sought. It is similar to Hvorslev's Equation (2.28) with (2.31), except that R_e appears in place of L. If R_e can be estimated, than K can be determined from the water level rise.

Values of R_e, expressed in terms of $\ln (R_e / r_w)$, were determined by Bouwer and Rice [12] with an electrical resistance network analog for different values of r_w, L, D, and H (see Figure 2.14). The results from these analog simulations provided the following empirical equation:

$$\ln (R_e / r_w) = \left(\frac{1.1}{\ln (D/r_w)} + \frac{A + B \ln [(H - D)/r_w]}{L/r_w} \right)^{-1} \tag{2.35}$$

In Equation (2.35), A and B are dimensionless coefficients that are functions of L/r_w, as shown in Figure 2.15. If $H \gg D$, Bouwer and Rice find that an increase in D has no measurable effect on $\ln (R_e / r_w)$. The analog results indicated that the effective upper limited of $\ln [(H - D)/r_w]$ is 6. Thus if H is very large so that the calculated value of $\ln [(D - H)/r_w]$ is greater than 6, a value of 6 should still be used for the term $\ln[(H - D)/r_w]$ in Equation (2.35).

If $D = H$, then the term $\ln [(H - D)/r_w]$ cannot be used. The analog results of Bouwer and Rice indicated that for this condition, which is the case of a fully penetrating well, Equation (2.35) should be modified to

$$\ln \left(\frac{R_e}{r_w} \right) = \left(\frac{1.1}{\ln (D/r_w)} + \frac{C}{L/r_2} \right)^{-1} \tag{2.36}$$

where C is a dimensionless parameter which is a function of L/r_w as shown in Figure 2.15.

Bouwer and Rice [12] suggest that Equations (2.35) and (2.36) yield values of $\ln(R_e/r_w)$ that are within 10% of the actual analog simulation values if $L > 0.4\,D$ and within 25% if $L \ll D$ (for example, $L = 0.1\,D$). Example 2.4 (after Bouwer and Rice [12]) shows an application of the Hvorslev and the Bouwer and Rice methods of slug test analysis.

Example 2.4

A slug test was performed on a cased well in the alluvial deposits of the Salt River bed west of Phoenix, Arizona. The static water table was at a depth of 3 m, $H = 80$ m, $D = 5.5$ m, $L = 4.46$ m, $r_c = 0.076$ m, and r_w was taken as 0.12 m to allow for development of the aquifer around the perforated portion of the casing. A solid cylinder with a volume equivalent to a 0.32-m change in water level in the well was placed below the water table. After equilibrium levels were again achieved, the cylinder was quickly removed and the water levels were determined from analysis of the output of a pressure transducer. The results are shown in Figure 2.16. The straight-line portion is the valid part of the readings. The actual y_0 value of 0.29 m indicated by the straight line is close to the theoretical value of 0.32 m calculated from the displacement volume of the cylinder.

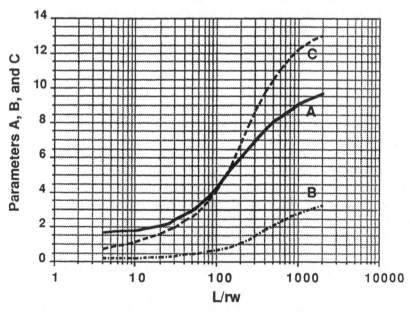

Figure 2.15 Curves relating coefficients A, B, and C to L/r_w (redrawn from Bouwer and Rice, 1976 [12]).

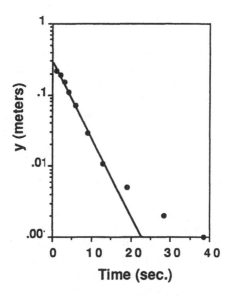

Figure 2.16 Data from a slug test in a permeable soil (from Bouwer and Rice [12]).

For this well, $L/r_w = 37$, so from Figure 2.15 one may estimate that $A = 2.7$ and $B = 0.45$. Substituting these values into Equation (2.35) and using the maximum value of 6 for $\ln[(H - D)/r_w]$ gives $\ln(R_e/r_w) = 2.31$. Using this value in Equation (2.34) gives

$$K = \frac{0.076^2 \times 2.31}{2 \times 4.56} \frac{1}{22.5} \ln \frac{.29}{.001} = 0.00037 \text{ m/s} = 32 \text{ m/d}$$

which agrees well with K values of 10 and 53 m/d obtained by other methods in this area.

For comparison, consider Hvorslev's method for this data. With the notation of Figure 2.12, one has $y_0 = H - H_0 = 0.29$ m, so that τ occurs when $y_\tau = H - h = 0.37 \times 0.29 = 0.107$ m. According to Figure 2.16 this occurs at a time of about 4 seconds. With Equation (2.31) one has

$$F = \frac{2 \times \pi \times 4.56}{\ln(4.56/0.076)} = 7.0$$

and from Equation (2.28) with $A = \pi r_c^2 = 0.018$ m², one obtains

$$K = \frac{0.018}{7.0 \times 4} = 0.00064 \text{ m/s} = 56 \text{ m/d}$$

which is on the same order of magnitude but significantly larger.

FLOW OF AIR THROUGH POROUS MEDIA

The flow of air through soil is of interest for a number of applications. Air may be used as the fluid in laboratory columns for measurement of permeability of a porous media. Also, the forced flow of air is very important in design of soil vapor extraction systems for remediation of unsaturated soils that have become contaminated by volatile organic substances. While bulk air flow is important for many other applications, the basic principles will be presented in terms of these applications.

For air flow it is still generally assumed that Darcy's law applies. Again, this implies that the driving force from pressure gradients and gravity is balanced by the viscous resistance force associated with the flow, and further that this viscous resistance force is proportional to the velocity to the first power. However, since the density of air is not constant, one cannot define a hydraulic head as is done for the flow of water. One can show that the general form of Darcy's law for the flow of air is the same as Equation (2.6), which is thus valid both for incompressible and compressible fluids. Under most conditions the force associated with pressure gradients is much larger than that due to gravity (the density of air is so small), and Darcy's law may be written in the approximate form

$$\mathbf{U} = -\frac{k}{\mu}\nabla p \qquad (2.37)$$

Equation (2.37) is exact for horizontal flow, and may usually be accepted in general for flow of air in porous media.

When developing the continuity equation for flow of soil air one may generally neglect compressibility of the porous matrix. In addition, if the transport process is isothermal then Boyle's law applies as an equation of state:

$$p = \varrho \, \frac{p_0}{\rho_0}$$

where p now must be the absolute pressure and p_0 and ϱ_0 are the absolute pressure and density at some reference or standard state (such as local atmospheric conditions). Under these conditions the continuity equation reduces to

$$n \, \frac{\partial p}{\partial t} = \nabla \cdot \left(p \, \frac{k}{\mu} \, \nabla p \right) \qquad (2.38)$$

Equation (2.38) is obviously nonlinear and only approximate solutions are available.

For some applications, especially those associated with soil vapor extraction systems, Equation (2.38) may be linearized by assuming that the pressure is equal to atmospheric plus a small deviation:

$$p = p_{atm} + p^*$$

where $p^* \ll p_{atm}$. Under these conditions the viscosity of a vapor is also nearly constant and if the intrinsic permeability is also taken as constant, then Equation (2.38) reduces to

$$\frac{n\mu}{kp_{atm}} \frac{\partial p^*}{\partial t} = \nabla^2 p^* \tag{2.39}$$

In Equation (2.39), p^* represents the deviation of the air pressure from atmospheric.

Equation (2.39) has the same basic form as Equation (2.12). This suggests that the time-dependent pressure response to a vapor extraction well may be estimated from the equivalent form of the Theis equation given by Johnson et al. [13].

$$p^* = \frac{Q}{4\pi b(k/\mu)} W\left(\frac{r^2 n\mu}{4kp_{atm}t}\right)$$

or from the equivalent form of the Jacob equation

$$p^* = \frac{Q}{4\pi b(k/\mu)} \ln\left(\frac{2.25\ kp_{atm}t}{r^2 n\mu}\right) \tag{2.40}$$

Equation (2.40) may be used to evaluate the subsurface characteristics for design of a soil vapor extraction system.

REFERENCES

1 de Marsily, G. 1986. *Quantitative Hydrogeology: Groundwater Hydrology for Engineers,* Academic Press, Orlando.
2 Meinzer, O. E., ed. 1942. *Hydrology,* Dover, New York.
3 Bird, R. B., W. E. Stewart and E. N. Lightfoot. 1960. *Transport Phenomena,* John Wiley, New York.
4 Theis, C. V. 1935. "The Relationship Between the Lowering of the Piezometric Surface and the Rate and Duration of Discharge of a Well Using Groundwater Storage," *Trans. Amer. Geophys. Union,* 2:519–524.
5 Freeze, R. A. and J. A. Cherry. 1979. *Groundwater,* Prentice-Hall, Englewood Cliffs.

6 Cooper, H. H., Jr. and C. E. Jacob. 1946. "A Generalized Graphical Method for Evaluating Formation Constants and Summarizing Well Field History," *Trans. Amer. Geophys. Union*, 27:526–534.

7 Kruseman, G. P. and N. A. de Ridder. 1970. *Analysis and Evaluation of Pumping Test Data,* Intern. Inst. Land Reclamation and Improvement Bull. 11, Wageningen, The Netherlands.

8 Walton, W. C. 1970. *Groundwater Resource Evaluation*, McGraw-Hill, New York.

9 Bouwer, H. 1978. *Groundwater Hydrology,* McGraw-Hill, New York.

10 Bouwer, H. and R. D. Jackson. 1974. "Determining Soil Properties," in *Drainage for Agriculture*, J. V. Schilfgaarde, ed., American Society of Agronomy, Agronomy No. 17, Madison, Wisconsin.

11 Hvorslev, M. J. 1951. "Time Lag and Soil Permeability in Ground-Water Observations," Bulletin No. 36, Waterways Experiment Station, Corps of Engineers, U.S. Navy, Vicksburg, Mississippi.

12 Bouwer, H. and R. C. Rice. 1976. "A Slug Test for Determining Hydraulic Conductivity and Unconfined Aquifers with Completely or Partially Penetrating Wells," *Water Resour. Res.*, 12(3):423–428.

13 Johnson, P. C., C. C. Stanley, M. W. Kemblowski, D. L. Byers and J. D. Kolthart. 1990. "A Practical Approach to the Design, Operation and Monitoring of In Situ Soil-Venting Systems," *Ground Water Monit. Rev.*

CHAPTER 3

Subsurface Contaminant Transport

W HEN a pollutant is released from a waste storage facility, as shown in Figure 3.1, it migrates downward through the unsaturated zone (or vadose zone) to the water table and then laterally in the direction of the hydraulic gradient in the saturated zone. Throughout this transport its fate is controlled by a myriad of physical, chemical, and biotic processes. These include the physical processes of advection, diffusion, dispersion, and capillarity, and the biotic and abiotic processes of bioaccumulation, degradation, immobilization, retardation, and volatilization. Quantification of these various processes at a field site is very difficult.

Assessment of subsurface fate and transport must address questions of source characterization (what is released, where, when, how much, etc.), vadose zone transport and processes, groundwater transport, and exposure and dose assessment.

PHYSICAL MECHANISMS OF SOLUTE TRANSPORT

There are three basic physical mechanisms by which miscible and immiscible pollutants are transported in the subsurface environment: advection, diffusion, and mechanical dispersion. Emphasis in this section is placed on soluble species, and only single phase flow is considered. Multiphase transport of immiscible pollutants will be considered in a later chapter. Advection refers to the soluble species being carried forward with the flow of subsurface water. As seen in Chapter 2, the volumetric flux of water, or the Darcy velocity, is given by

$$\mathbf{U} = -K \operatorname{grad} h = K\mathbf{I} \tag{3.1}$$

where h is the hydraulic head and $\mathbf{I}$ is the hydraulic gradient. For unsaturated flow, K is a function of the water content. The average speed of

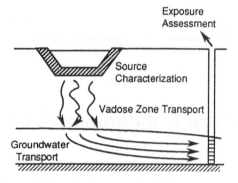

Figure 3.1 Scenario for exposure and dose assessment.

the water movement is given by the pore water velocity or seepage velocity, $\mathbf{v}$. The seepage and Darcy velocity are related through

$$\mathbf{v} = \frac{\mathbf{U}}{n_e} \tag{3.2}$$

for saturated flow where n_e is the kinematic or effective porosity, and

$$\mathbf{v} = \frac{\mathbf{U}}{\theta} \tag{3.3}$$

for unsaturated flow where θ is the volumetric water content. For most soils it is assumed that $n_e \cong n$ and no distinction is drawn between the effective porosity and the total porosity. An exception is drawn for clay soils and clay liners for waste disposal facilities, where the difference between the effective and total porosity may be significant.

For a solute concentration c, the advective flux past a bulk surface area is

$$\mathbf{U}c \tag{3.4}$$

With Darcy's law it is seen that the advective flux equals $K\mathbf{I}c$. Thus the bulk transport is proportional to the medium hydraulic conductivity, the energy gradient, and the local concentration.

While advection is associated with the bulk macroscopic groundwater movement, diffusion is a molecular-based phenomenon. If one could see the individual molecules, one would note the continual movement of each molecule and of one molecule relative to another. When these random molecular movements occur in a field with a concentration gradient, there is a net movement of the species toward regions of lower concentration. This is the process of diffusion. According to Fick's law of diffusion (Fick's first law), the diffusive mass flux for a saturated porous medium is

$$- nD_s \text{ grad } c \qquad (3.5)$$

where D_s is the apparent diffusion coefficient in soil for the chemical species. D_s is smaller than the molecular diffusion coefficient because the solute is confined to moving along a tortuous path through the pore space. For unsaturated flow, the tortuosity increases with decreasing water content and D_s is smaller still. For saturated media, values of the apparent diffusion coefficient are on the order of 10^{-4} m²/d.

The third mechanism of pollutant transport is associated both with bulk fluid movement and with the presence of the porous media with its complex, intertwining pore space. Fluid particles that are at one time close together tend to move apart because of at least three physical mechanisms. First, the particles nearest the walls of the pore channel move slower than those near the channel center. Second, the variations of pore dimensions along the pore axes cause the particles to move at different relative speeds. Third, adjacent particles in one channel can follow different streamlines that lead to different channels. These particles may later come together in another channel or they may continue to move further apart. When these mechanisms occur in the presence of a concentration gradient the resulting transport relative to the bulk water movement is referred to as mechanical dispersion. In a very definite sense dispersion occurs because of our inability to follow the details of pore to pore scale groundwater movement.

Statistically, advection refers to the average rate of movement while mechanical dispersion refers to the deviation from the mean. Also statistically, dispersion will be scale dependent. The further a particle moves in the subsurface, the greater the range of heterogeneities of hydraulic conductivity it will experience. For example, the particle may either start off in a sand or in a clay. For short distances of movement it will remain in the same type of material it started off in and the dispersion coefficient will be characteristic of that material. However, as it moves further from its initial point it may move from sand to clay to sand, etc., with each unit having its own characteristic velocity. Considering two particles, it is apparent that the expected deviation of their locations from the mean position will increase more through this actual heterogeneous system than it would through an idealized homogeneous system.

The mechanical dispersion mass flux is usually modeled as a Fickian type of process. However, field and laboratory experience suggests that the mixing is greater in the direction of flow than transverse to this direction. Also, for uniform flow the dispersion coefficient (mixing coefficient) is found to be proportional to the flow rate. These observations suggest that the dispersive flux in the direction of flow (the longitudinal dispersion) may be modeled as

$$-a_L \mathbf{U} \cdot \operatorname{grad} c \qquad (3.6)$$

while the flux in the direction transverse to the mean flow (the transverse dispersion) is given by

$$-a_T \mathbf{U} \cdot \operatorname{grad} c \qquad (3.7)$$

One of the goals of the theory of dispersion is to generalize these relationships for nonuniform flow fields. The coefficients a_L and a_T are referred to as the longitudinal and transverse dispersivities (units of length). In laboratory experiments, the longitudinal dispersivity is usually found to be five to twenty times larger than the transverse dispersivity. In the laboratory, a_L has been found to vary from 0.1 to 10 mm. In the field, the dispersivities are sometimes measured through single and multiple well tracer tests. More often, however, what is usually done is that measured field concentrations are simulated with mathematical models and the coefficients adjusted to get an adequate match. The values found in this fashion are usually much larger than laboratory values. Recent literature has shown field values of a_L to vary from 1 to 100 m or larger. These values of a_L are larger than laboratory values by a factor of up to 10^5, suggesting that dispersion plays a different role in the field than in the laboratory.

In practice one usually combines the coefficients of diffusion and mechanical dispersion into a single hydrodynamic dispersion coefficient. Because of mechanical dispersion, this new coefficient will depend upon direction (mixing is greater along the direction of flow as compared with the transverse direction) and the hydrodynamic dispersion coefficient is actually a second order symmetric tensor.

CONTINUITY PRINCIPLES

The fundamental equation of pollutant transport is the *conservation of mass equation*. This states that for an arbitrary region of the aquifer:

net rate of mass   net mass flux   increase in mass due to
increase within   =   into region   +   biotic/abiotic reactions within
the region   the region

$$(3.8)$$

The mass increase term represents the total mass per bulk volume, including both the sorbed mass and that in solution. For a saturated medium the mass density, m, which is the mass per bulk volume, may be represented as

$$m = nc + (1 - n)\varrho_s q \qquad (3.9)$$

where m is the bulk concentration (mg/L), ϱ_s is the soil density (usually 2.6 to 2.7 gm/cc), and q is the sorbed concentration (units of mass or activity sorbed per mass of soil). The product $(1 - n)\varrho_s$ is the bulk density, ϱ_b, of the soil. In Equation (3.9), q and c are related through a sorption isotherm if equilibrium conditions hold, or by way of mass transfer rate terms of kinetic models of sorption are appropriate.

The net flux term in Equation (3.8) includes advective, diffusive, and dispersive mass transport. The mass flux vector, $\mathbf{J}$, is the mass crossing a unit area per unit time. The reaction term includes radioactive decay, biodegradation of organic pollutants, precipitation and redox chemical reactions that may immobilize a pollutant, and others, and may be represented as the symbol S^+. S^+ is the source strength and has units of mass per unit volume per unit time.

For the arbitrary control volume shown in Figure 3.2, the conservation of mass equation takes the form

$$\frac{d}{dt} \iiint m\,dV = - \iint \mathbf{J} \cdot \mathbf{n}\,dA + \iiint S^+ dV \qquad (3.10)$$

where the first term represents the time rate of increase in the total mass within the control volume, the second term is the net flux of mass into the volume across the control surface with $\mathbf{n}$ the outward unit normal vector to the control surface, and the last term is the mass increase due to sources located within the volume.

Equation (3.10) is an integral form of the continuity equation. Since the

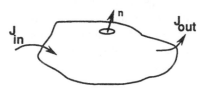

Figure 3.2 Control volume for mass balance.

control volume is arbitrary, one may also write the continuity equation in the form

$$\frac{\partial m}{\partial t} + \text{div } (\mathbf{J}) = S^+ \tag{3.11}$$

This is the most general form of the continuity equation and serves as the starting point for most further investigations of subsurface fate and transport. For most applications one works with a form of the general continuity equation that is simplified in one fashion or another. These simplifications involve making further assumptions as to how to model the various processes that are of interest.

Nonpolar organic compounds in groundwater are found to be sorbed by the medium on existing solid organic matter present in the porous medium. This sorption is due primarily to hydrophobic interactions resulting in weak, nonspecific sorption forces. When the organic compounds are present in trace concentrations, linear sorption isotherms are often observed

$$q = K_d c \tag{3.12}$$

where K_d is the distribution or partition coefficient for the chemical species (L/kg). The distribution coefficient is found to be a function of the hydrophobic character of the organic compound and the amount of organic matter present and may be written

$$K_d = K_{oc} f_{oc} \tag{3.13}$$

where K_{oc} is the organic carbon partition coefficient and f_{oc} is the fraction of organic carbon within the soil matrix. Sorption partition coefficients, indexed to organic carbon (K_{oc}) are relatively invariant for natural sorbents, and K_{oc}'s can be estimated from other physical properties of pollutants such as aqueous solubility or octanol/water partition coefficients. Equation (3.13) is valid only for $f_{oc} > 0.001$. Otherwise, sorption of organic compounds on nonorganic solids (clays and mineral surfaces) can become significant. Also, the linear isotherm model is valid only if the solute concentration remains below one-half of the solubility limit of the compound. With the linear sorption isotherm and assuming local equilibrium, the expression for the bulk concentration can be written

$$m = (n \times \varrho_b K_d)c \tag{3.14}$$

Equation (3.14) expresses the bulk concentration in terms of the concentration in solution.

Losses due to biotic processes are often modeled as either first- or zero-order decay. Assuming that the source term is actually represented as a first-order equation with an apparent or effective rate constant, λ, then

$$S^+ = -\lambda m \qquad (3.15)$$

This relationship states that the loss rate is proportional to the total mass present. However, research has shown that biodegradation occurs only if the solute is present in the aqueous phase where the microorganisms may directly attack it. As long as linear partitioning relationships hold, then Equation (3.15) remains true as an effective rate constant, but this does not actually represent the loss rate from the aqueous phase.

Finally, we may represent the mass flux vector in terms of its advective and dispersive components as

$$\mathbf{J} = \mathbf{U}c - n\mathbf{D_{hd}} \cdot \text{grad} \ (c) \qquad (3.16)$$

where $\mathbf{D_{hd}}$ is the hydrodynamic dispersion tensor.

Substitution of Equations (3.14), (3.15), and (3.16) in the general continuity Equation (3.11) gives

$$(n + \varrho_b K_d) \frac{\partial c}{\partial t} + \text{div} \ (\mathbf{U}c) + \lambda(n + \varrho_b K_d)c$$

$$= \text{div} \ [n\mathbf{D_{hd}} \cdot \text{grad} \ (c)] \qquad (3.17)$$

If the porosity is constant and approximately equal to the kinematic porosity then one may introduce the *retardation factor* as

$$R = 1 + \frac{\varrho_b}{n} K_d \qquad (3.18)$$

The physical significance of the retardation factor is that it measures how much slower the solute migrates than water because the solute spends part of its time sorbed on the soil matrix and immobile. Thus a retardation factor of ten means the average speed of the solute is ten times slower than that of the water. In addition, if the flow field is steady and there are no volumetric sources or sinks (leakage, infiltration, evaporation, etc.) then Equation (3.17) may be written as

$$R \frac{\partial c}{\partial t} + \mathbf{v} \cdot \text{grad}\ (c) + \lambda Rc = \text{div}\ [\mathbf{D_{hd}} \cdot \text{grad}\ (c)] \qquad (3.19)$$

Equation (3.19) is the form of the continuity equation that is used for most analytical models in one, two, and three dimensions with the added assumption that the velocity field is uniform. In particular, the one-dimensional form of the continuity equation is

$$R \frac{\partial c}{\partial t} + v \frac{\partial c}{\partial x} - D \frac{\partial^2 c}{\partial x^2} + \lambda Rc = 0 \qquad (3.20)$$

The first term in Equation (3.20) gives the concentration change at a given location (including the mass in solution as well as the sorbed mass). The second term accounts for the change in concentration associated with advection. The third term accounts for the concentration change associated with mixing or diffusion, and the last term is the sink term which is modeled as first-order decay.

SOLUTE TRANSPORT MODELS

Mathematical models are used to simulate the subsurface transport of chemicals. Both analytical and numerical models are used. The numerical models are the most general since they may be tailored to address site-specific conditions. However, these numerical models require a significant data base, which may not be available in the initial phases of a site investigation. Analytical models require simplifying assumptions, but they are much more computationally efficient and need less specific data. They are especially useful for screening calculations. This section is concerned with simple calculations from analytical models for column experiments, chemical spills, and chemical plumes from continuous releases of contaminants.

First consider the transport of a conservative tracer. This is a substance that moves through the porous media without interacting with the matrix or undergoing chemical or biotic transformations. Understanding the transport of a conservative substance is the first step toward understanding the fate and transport of hazardous and radioactive chemicals, which may be influenced by a wide range of processes. The advection-dispersion equation for a conservative species takes the form

$$\frac{\partial c}{\partial t} + \mathbf{v} \cdot \text{grad } (c) = \text{div } [\mathbf{D}_{hd} \cdot \text{grad } (c)] \tag{3.21}$$

This is the form of the transport equation that is usually applied. In order to obtain an analytical solution one generally must assume that the flow is uniform in a particular direction, say in the x-direction. In this case Equation (3.21) becomes

$$\frac{\partial c}{\partial t} + v_x \frac{\partial c}{\partial x} = D_{xx} \frac{\partial^2 c}{\partial x^2} + D_{yy} \frac{\partial^2 c}{\partial y^2} + D_{zz} \frac{\partial^2 c}{\partial z^2} \tag{3.22}$$

In particular, for a one-dimensional problem one has

$$\frac{\partial c}{\partial t} + v \frac{\partial c}{\partial x} = D \frac{\partial^2 c}{\partial x^2} \tag{3.23}$$

where D is the combined mechanical dispersion and diffusion coefficient. These processes cannot be distinguished in a one-dimensional flow.

COLUMN EXPERIMENTS

It is of obvious interest to be able to estimate the dispersion coefficient for a porous medium. The simplest means for doing this, and the method that has been used most often, is through column experiments. During these experiments a tracer is passed through a column, generally upward to help in the displacement of entrapped air, and the effluent concentration is measured as a function of time. The problem is modeled mathematically, and the experimental data is fit against the mathematical solution to the problem to find the parameter values that provide the best fit between the theory and observed data. For column experiments with ideal tracers, the transport equation is Equation (3.23). The choice of boundary conditions is fairly difficult. At the inlet end of the column, the best approach is to provide a mass balance for the source reservoir as the boundary condition. Unfortunately, this is very difficult to deal with mathematically (though a few solutions are known), and the usual approach is to specify a given concentration at the influent end of the column or to match the flux out of the reservoir with that into the porous medium. The effluent end presents even greater problems because it is neither a constant flux nor a constant concentration boundary. The favored approach is to treat the effluent end as a "free outfall," specifying neither the concentration nor the flux. This is done by specifying the concentration at a downstream point, namely infinity. The effluent concentration may then be found by evaluating the predicted concentration at the end of the column as a function of time, $c(L,t)$, where L is the length of the column. For the initial condition we assume that the column is initially free of the tracer.

For these boundary and initial conditions there are a number of available solutions that may be used to analyze column experiment data. Perhaps the simplest and most useful is given by the approximate equation

$$\frac{c}{c_0} \cong \frac{1}{2} \, \text{erfc} \left(\frac{x - vt}{\sqrt{4Dt}} \right) \tag{3.24}$$

where erfc( ) represents the complementary error function. For the typical range of parameters, the more complete solutions reduce basically to this form. Figure 3.3 shows Equation (3.24) for a case with a column of length $L = 0.5$ m, an average pore water velocity of $v = 1.0$ m/d, and dispersion coefficients of 10^{-3}, 10^{-2}, and 0.1 m²/d, respectively. These correspond to dispersivities of 0.001, 0.01, and 0.1 m. Dispersivity values on the 1 mm level are common for laboratory column experiments, while values on the order of 0.1 m or larger are often observed in the field. It is apparent that the amount of mixing that appears in the field is much greater than that found in the laboratory, and this is associated with the

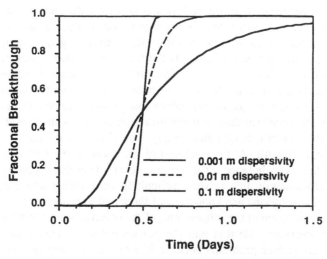

Figure 3.3 Breakthrough curves from column experiments with $L = 0.5$ m and $v = 1$ m/d.

much wider range of heterogeneities found in natural geologic deposits. For a reactive tracer with retardation factor R, the equivalent form of Equation (3.24) is

$$\frac{c}{c_0} \cong \frac{1}{2}\, \mathrm{erfc}\left(\frac{x - \dfrac{vt}{R}}{\sqrt{\dfrac{4Dt}{R}}}\right) \tag{3.25}$$

Comparison of Equations (3.24) and (3.25) shows that the effect of retardation without any decay is to slow the transport process down. For example, if $R = 2$, then the curves in Figure 3.3 would remain unchanged if the times were all doubled (i.e., day 1 becomes day 2, etc.).

SPILL MODEL IN TWO AND THREE DIMENSIONS

Contaminant releases to the subsurface may occur over short periods of time, or there may be a slow continual release over a long time period. The first of these cases may be modeled by assuming that the release is instantaneous, while the second leads to plume models. If a volume V_0 containing a chemical at a concentration c_0 is released to the water table over a short duration, the chemical slug is subsequently transported downgradient as it spreads in the direction of flow, transverse horizontally and transverse across the thickness of the aquifer. Near the source the spreading is three-dimensional, and if the flow is uniform in the x-direction at a velocity v, then the concentration is given by

$$c(x,y,z,t) = \frac{2c_0 V_0 e^{-\lambda t}\exp\left(-\dfrac{\left(x - \dfrac{vt}{R}\right)^2}{\dfrac{4D_{xx}t}{R}} - \dfrac{y^2}{\dfrac{4D_{yy}t}{R}} - \dfrac{z^2}{\dfrac{4D_{zz}t}{R}}\right)}{nR\sqrt{\left(\dfrac{4\pi D_{xx}t}{R}\right)\left(\dfrac{4\pi D_{yy}t}{R}\right)\left(\dfrac{4\pi D_{zz}t}{R}\right)}} \tag{3.26}$$

According to Equation (3.26) the maximum concentration occurs at the location $x = vt/R$ and $y = z = 0$, and is given by

$$c_{max} = \frac{2c_0 V_0 e^{-\lambda t}}{nR\sqrt{\left(\dfrac{4\pi D_{xx}t}{R}\right)\left(\dfrac{4\pi D_{yy}t}{R}\right)\left(\dfrac{4\pi D_{zz}t}{R}\right)}} \tag{3.27}$$

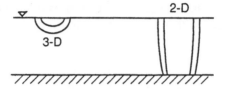

Figure 3.4 Schematic view of spreading of point release model showing the near-field and far-field distributions.

The main feature of interest in Equation (3.27) is that near the source, the maximum concentration decreases as

$$c_{max} \sim \frac{1}{t^{3/2}} \tag{3.28}$$

or since the distance of migration of the centroid of the contaminated mass is given by $L = vt/R$, the concentration decreases as $L^{-3/2}$.

Far from the source the contaminant will have spread over the entire thickness of the aquifer, and the continued transport may be represented by a two-dimensional model as shown in Figure 3.4. The solution given by Equation (3.26) transforms into

$$c(x,y,t) = \frac{c_0 V_0 e^{-\lambda t} \exp\left(-\dfrac{\left(x - \dfrac{vt}{R}\right)^2}{\dfrac{4D_{xx}t}{R}} - \dfrac{y^2}{\dfrac{4D_{yy}t}{R}}\right)}{4\pi nbt\sqrt{D_{xx}D_{yy}}} \tag{3.29}$$

where b is the thickness of the aquifer. This equation shows that far from the source the maximum concentration decreases as $1/t$ or $1/L$, which is much slower than that found near the source. This is because mixing is confined to a two-dimensional domain far from the source, while near the spill location, mixing can occur in the vertical direction as well.

It is not difficult to find the solution for the intermediate domain where neither Equation (3.26) nor (3.29) is valid. This solution shows that Equation (3.26) is valid for

$$L < \frac{0.036 \, vb^2}{D_{zz}} \tag{3.30}$$

while Equation (3.29) is valid for

$$L > \frac{0.8 \, vb^2}{D_{zz}} \qquad (3.31)$$

With an aquifer of thickness $b = 10$ m and a vertical dispersivity of $a_v = 5$ cm, Equations (3.30) and (3.31) correspond to distances of 72 m and 1600 m, respectively.

CONTAMINANT PLUME MODEL

If a contaminant is released at a constant rate from a source for a long enough period of time, then a steady concentration distribution in the shape of a contaminant plume is developed. The steady plume models are useful in endangerment studies because they provide the maximum likely concentration that may be observed at a given location. The discussion here considers a very simple plume model, which is important because all other models reduce to this form in the far field. This model assumes that the contaminant has been mixed over the aquifer thickness (or that the exposure well is screened over the aquifer thickness) and that the release occurs at a localized point at a rate $\dot{m}$ (kg/day, for example). The concentration distribution is given by

$$c(x,y) = \frac{\dot{m} \exp\left(-\dfrac{\lambda R x}{v}\right)}{nb\sqrt{4\pi v D_{yy} x}} \exp\left(-\dfrac{y^2}{4 \dfrac{D_{yy}}{v} x}\right) \qquad (3.32)$$

The following example will consider the effect of parameter variability. In order to examine the influence of the various parameters appearing in Equation (3.32), consider a problem with the following base characteristics: $n = 0.3$, $b = 5$ m, $\dot{m} = 5$ kg/d, $v = 0.2$ m/d, $R = 1$, $D_{yy} = a_T v$, $a_T = 1$ m, and $\lambda = 0$ d^{-1}. Figure 3.5 shows the effect on the plume concentration of varying the velocity from 0.02 m/d to 2 m/d. A one order-of-magnitude increase in v results in an equivalent decrease in the concentration. This occurs because the mass flow to the aquifer ends up mixing with a greater amount of groundwater when the velocity increases; thus, the effect seen in Figure 3.5 is entirely due to dilution of the source.

Figure 3.6 shows the effects of increasing the transverse dispersivity from 0.1 m to 10 m. The corresponding change in concentration is not as large as in Figure 3.5 because the dispersivity appears under the square

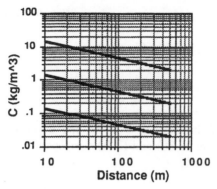

Figure 3.5 Centerline concentration from simple plume model for seepage velocities of 0.02 m/d, 0.2 m/d, and 2 m/d, respectively.

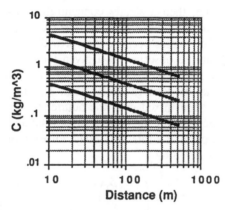

Figure 3.6 Centerline concentrations predicted the simple plume model with dispersivities of 0.1 m, 1 m, and 10 m, respectively.

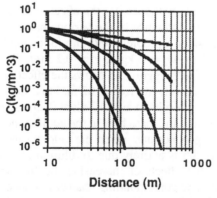

Figure 3.7 Centerline concentrations predicted by the simple plume model for cases with no decay and with half-lives of 400 days, 100 days, and 30 days, respectively.

root in the denominator of Equation (3.32). Here we see that a two orders-of-magnitude increase in a_T results in a one order-of-magnitude decrease in c.

Finally, Figure 3.7 shows the effect of decay. The upper curve corresponds to the case with no decay, while the lower curves correspond to half-lives of 400 days, 100 days, and 30 days, respectively. It is apparent that with even moderate levels of first-order losses, the concentrations decrease rapidly with distance from the source.

CHAPTER 4

Bioremediation of Contaminated Soils and Sludges

BACKGROUND

REGULATORY CONCERN

UNTIL the early 1980s, there were few specific regulatory requirements for the remediation of contaminated soils and of industrial sludges contained in pits, ponds, and lagoons. Factors that have changed attitudes about these soils and sludges include:

- Groundwater pollution from such sites created public interest and demonstrated how costly cleanups could be.
- Investigations indicated that a large number of leaky underground storage tanks existed. This focused local as well as state and federal attention on hydrocarbon spill situations.
- There is increasing regulatory concern that some petroleum hydrocarbons contain constituents now considered to be toxic.
- The implementation of Resource Conservation Recovery Act (RCRA) requirements compel organizations to evaluate potential contamination at their sites.

As a result, the need for remedial action has resulted in regulations that require action to reduce these hazards in a cost-effective and environmentally sound manner. The existing regulations result from several legislative mandates:

- the Safe Drinking Water Act (SDWA) – protecting and restoring aquifers
- the Resource Conservation and Recovery Act (RCRA) – destroying or detoxifying hazardous wastes
- the Comprehensive Environmental Response, Compensation, and Liability Act (CERCLA) – restoring Superfund sites and decontaminating soils

- the Federal Insecticide, Fungicide, and Rodenticide Act (FIFRA)—waste pesticide disposal

During the mid-1980s, a common remediation approach for contaminated soils and sludges was to excavate and transport the soils and sludges to an approved disposal site since this usually was the least costly and most expeditious solution. Over time, this solution became less attractive to industry, regulators, and the public. Available landfill space was decreasing quickly, many sites closed, expansion of others became increasingly difficult, and the excavate, transport, and dispose approach was recognized as not a permanent solution.

In 1986, CERCLA was amended to encourage the use of remediation technologies that would result in permanent solutions, i.e., reduce the toxicity and immobilize the contaminants in the soils and sludges. Thus, the regulatory and technical challenge now is to use cost-effective control technologies that can treat complex chemical mixtures in contaminated soils and sludges and thereby reduce the threat to human health and the environment. Bioremediation can be such a technology.

BIOREMEDIATION CONCEPTS

Remediation is the process of removing or reducing the mobility and/or toxicity of the contaminants of concern at a site. Remediation processes applicable to saturated soils and remediation of groundwaters are not discussed in this document. The focus of this document is on bioremediation processes that can be used for unsaturated surface soils and the sludges contained in pits, ponds, and lagoons that are constructed above the water table. These are bioprocesses in which the dominant electron acceptor is oxygen supplied directly from the atmosphere or by mechanical aerators in surface pits, ponds, lagoons, or tanks. With subsurface treatment processes, the electron acceptor is supplied by perfusing the contaminated soil with air or water. In-situ bioremediation processes are subsurface treatment processes and are not described in this document.

The term bioremediation is used to mean biological treatment processes in unsaturated surface soils that result in the degradation of industrial-based organics to simpler organic molecules and ultimately carbon dioxide and water. In addition to biodegradation, immobilization and volatilization of some components may occur during the treatment process.

Bioremediation is a managed, demonstrated active treatment process that uses microorganisms to degrade and transform organic chemicals in contaminated soil, sludges, and residues. As such, bioremediation processes can be considered as source control, pollution prevention, and risk reduction processes that can reduce or eliminate groundwater contamination and thereby reduce the need for costly and long-term groundwater treatment processes.

Bioremediation processes have resulted from the application of knowledge from microbiology, biochemistry, environmental engineering, and chemical engineering. The fundamentals of biological treatment, as applied to site-specific conditions, are the key to the proper design and separation of the bioremediation processes. In such processes, established scientific and engineering principles are used to maintain satisfactory conditions for microbial degradation and loss of organics.

The advantages of bioremediation processes are that the processes:

- are used where the problem is located
- do not require transporting large quantities of contaminated material offsite
- eliminate the problem rather than moving it somewhere else
- minimize long-term liability
- are ecologically sound and an extension of natural processes
- generally are cost-effective and competitive with other decontamination technologies for organics

Bioremediation commonly is part of a total remediation system for a site and can be used with other technologies to remediate a site and lower the overall cost of site cleanup.

Many bioremediation processes exist. Choosing the appropriate process is a function of the remediation goals to be achieved, the physical and chemical characteristics of the material to be treated, the environmental conditions that are created, the materials handling and equipment requirements, and the overall economics. The potential processes include:

- solid phase processes – such as composting and land treatment
- slurry phase processes – such as liquid solids systems in impoundments or bioreactors

Bioremediation processes have been used in the United States at Superfund sites. An evaluation of the records of decisions (RODs) that have been made for Superfund site remediations through 1989 indicates that bioremediation has been recommended for source control at over twenty sites. The types of bioremediation processes that are used at these sites are noted in Table 4.1.

The chemicals treated at these sites primarily have been polyaromatic hydrocarbons (PAH), volatile organics (benzene, toluene, ethylbenzene, and xylene, collectively noted as BTEX), pentachlorophenol, and phenols. Other chlorinated and nonchlorinated organics were present at some sites.

The factors that determine the feasibility of bioremediation and the appropriate bioremediation technology to consider for a particular site include:

TABLE 4.1. **Bioremediation Processes Recommended for Contaminated Soils and Sludges at Superfund Sites.***

Type of Process	Number of Times Recommended
Excavation followed by land treatment	11
Excavation with on-site treatment	3
In-situ treatment	4
Lagoon aeration	3
To be determined and other	2

*From review of Superfund RODs issued through 1989 (U.S.EPA, 1990) [1]; some RODs specify multiple remedies.

- type and quantity of hydrocarbons present in the contaminated soils and sludges
- volume of material requiring remediation
- depth to groundwater and quality of groundwater
- regulatory acceptance of the technology
- ultimate disposal of the residues from the bioremediation process
- time and costs to utilize a technology
- future site use

The characteristics of the material to be remediated (Table 4.2) also are important to the applicability of specific bioremediation processes.

TABLE 4.2. **Soil and Sludge Characteristics that Can Affect Feasibility of Bioremediation Processes.**

- *Solubility*—the measure of the potential for the constituent to become dissolved in water. Solubility affects mobility, leachability, availability for biodegradation, and the ultimate fate of the constituent.
- *Vapor pressure* (volatility)—the measure of the potential for the constituent to evaporate readily at standard temperature and pressure. The higher the vapor pressure, the more volatile the compound. Volatility impacts the level of off-gas control required for biological liquid/solids treatment and other bioremediation processes.
- *Viscosity*—the resistance of a fluid to continuous deformation when subjected to a shear stress. High viscosity constituents tend to fill the soil pore spaces and reduce the effectiveness of technologies which require the flow of air and water.
- *Degradation* (half-lives)—time required to reduce the concentration of the constituent in the soil to one-half its initial concentration. This provides a measure of the relative ability of a constituent to be degraded.
- *Toxicity*—toxic conditions can inhibit or not allow organics to be biodegraded.
- *Chemical properties*—other properties such as chemical structure of organics, pH, water-holding capacity of soils, temperature, contaminant sorption characteristics, and oxidation status (aerobic or anaerobic) can affect the amount and rate of degradation and immobilization that will occur.

PROCESS FUNDAMENTALS

OVERVIEW

The microbial degradation of organic compounds has been recognized for centuries as an efficient and relatively inexpensive method to treat organic compounds. Biotreatment processes such as composting for sludges and organic refuse, the activated sludge and trickling filter processes for wastewater, and anaerobic digestion for manures and organic sludges have been used for many decades. However, the application of biological processes to degrade and detoxify industrial organics in soils and sludges is not common. This is due primarily to a lack of understanding of: (1) the scientific and technical principles involved and (2) how those principles can be used to design and operate bioremediation processes. This section presents the pertinent principles and factors and discusses how they can be used and applied to such processes.

Bioremediation of a soil or sludge containing organics is accomplished by microbial degradation of specific organic compounds. Organisms in the soil require energy for food and growth. Microorganisms obtain this energy through the metabolic degradation of organic compounds. Thus, bioremediation involves the microbial breakdown of organic chemicals as a food and energy source. Generally microorganisms that currently exist in soil can degrade the organics in contaminated soils, assuming that the environmental conditions are suitable and nontoxic conditions exist. Microorganisms have evolved catalytic systems (enzymes) that degrade naturally occurring compounds present in the biosphere. However, contaminated soils and sludges can contain man-made organics that are difficult to degrade. With such chemicals, the natural microbial enzyme systems may have to adapt and acclimate to the chemical before degradation can occur. Acclimation results in an increase in the biodegradation rate of a chemical after the microbial community is exposed to the chemical for some period of time. A summary of the basic concepts that relate to bioremediation are noted in Table 4.3 and the factors that affect the performance of bioremediation processes are noted in Table 4.4.

The microbiological degradation of organics transforms elements from the organic to the inorganic state. The transformation of organic carbon to inorganic carbon (CO_2) is accomplished through enzymatic oxidation, with molecular oxygen involved as a terminal electron acceptor (aerobic metabolism). This also can occur if the final electron acceptor is something other than molecular oxygen, such as sulfate (SO_4) or nitrate (NO_3) (anaerobic metabolism). Aerobic degradation results in the production of CO_2; anaerobic degradation results in both methane (CH_4) and CO_2 production. Aerobic biological treatment produces innocuous end products — usually CO_2 and microbial biomass.

TABLE 4.3. **Basic Concepts of Bioremediation.**

- Bioremediation is a source control, risk reduction, and pollution prevention process.
- Bioremediation processes reduce the toxicity and migration potential of organic compounds.
- Biodegradation occurs in a wide variety of environments and both solid phase and slurry phase processes can be used for bioremediation.
- Organic compounds are microbially converted to simpler compounds.
- Microorganisms obtain the energy requirements for growth and maintenance from the compounds they degrade.
- Microbial enzymes evolved for the degradation of naturally occurring organics can be acclimated for the bioremediation of many organics in contaminated soils and sludges.
- Suitable environmental conditions are necessary for the successful bioremediation of contaminated soils and sludges.

TABLE 4.4. **Factors Affecting Bioremediation Processes.***

Factor	Comment
Microorganisms	Natural organisms are satisfactory; acclimation may be necessary; suitable environmental conditions need to be provided
Toxicity	Nontoxic conditions are needed
Available soil water	25–85% of water-holding capacity desirable for solid phase systems
Oxygen (O_2)	Aerobic conditions desired
Electron acceptors	Under aerobic conditions, O_2 is the terminal electron acceptor; when O_2 is not available, NO_3^-, Fe^{3+}, Mn^{2+} and SO_4^{2-} can act as electron acceptors
pH	5.5–8.5 for optimum degradation
Nutrients	Nitrogen, phosphorus, and other nutrients are needed for microbial growth
Temperature	Degradation rates are affected by temperature
Water solubility	The water solubility of compounds in contaminated soil can affect degradation
Sorption	Many organic compounds are strongly sorbed to the organic matter in soil
Volatilization	Chemicals can be lost by volatilization
Loss rates	Total loss rates commonly are reported; first-order rates usually used to describe losses

*From References [2] and [3].

58

In the soil, a constituent may not be completely degraded, but transformed to intermediate product(s). The goal of bioremediation processes is the detoxification of a parent compound to a product or products that are no longer hazardous. Thus, degradation may result in detoxification without complete mineralization.

If bioremediation can occur and is effective, why do some organics, such as those in contaminated soils and sludges, persist in the environment? This is a reasonable question that is answered by identifying the factors that affect microorganisms and therefore bioremediation processes. Even biodegradable organics may persist if adverse factors and nonoptimum conditions exist. The factors that may prevent microbial degradation and bioremediation include:

- chemical concentrations that are toxic to microorganisms
- inadequate type or numbers of microorganisms, such as due to toxic conditions
- conditions too acid or alkaline
- lack of nutrients such as nitrogen, phosphorus, potassium, sulfur, or trace elements (many organic chemicals, for example, are not nutritionally balanced)
- unfavorable moisture conditions (too wet or too dry)
- lack of oxygen or other electron acceptors

Bioremediation processes are biological treatment processes that improve or stimulate the metabolic capabilities of microbial populations to degrade organic residues. Therefore, it is important to understand those conditions and reactions so that bioremediation processes can be successful. With such knowledge, it is possible to modify the nonoptimum conditions so that microbial degradation can occur. For example, if there is inadequate nitrogen or phosphorus, such nutrients can be added to assure satisfactory microbial degradation. If the residues are too toxic, addition of other chemicals or uncontaminated soil may reduce the toxicity to the point that microbial degradation can occur. If inadequate types or numbers of microorganisms are present, acclimated organisms can be added.

The factors that affect bioremediation processes were identified in Table 4.4. The following sections provide details related to those factors.

MICROORGANISMS

Surface soils contain large numbers of microorganisms that include aerobic and anaerobic bacteria, fungi, actinomycetes, and protozoa as well as earthworms and higher forms of life. Over one million organisms can be present in one gram of agricultural surface soil. These organisms are capable of degrading most natural and synthetic organics that are in a soil.

However, they can accomplish the degradation only if: (1) nontoxic conditions exist, (2) the organisms have or can develop the enzyme systems capable of degrading the organic compound, and (3) other environmental conditions such as pH, nutrients, oxygen, temperature, and water are adequate.

The actual degradation that occurs in a bioremediation process is a result of the mixed microbial population that exists. One group of microorganisms may partially metabolize a compound and furnish a suitable substrate for another group of microorganisms. If an organic is biodegradable and environmental conditions are suitable, the natural organisms can adapt to degrade the organic compound. Specially developed microbial cultures have not been observed to be needed or successful in bioremediation systems.

TOXICITY

Many individuals who consider remediation options for contaminated soil assume that bioremediation is not feasible because chemicals identified as toxic or hazardous are in the soil. However, this assumption is not correct. Many organic chemicals in hazardous wastes can be biodegraded if the contaminated soil or sludge is not toxic to the microorganisms and if other environmental factors, such as pH, nutrients, and oxygen, are suitable.

In considering the feasibility of bioremediation processes, the relative toxicity of the media to be remediated should be determined. If the soil or sludge is found to be toxic to microorganisms, either steps should be taken to reduce the toxicity or another remediation process should be considered.

The usual procedures to quantify toxicity of a chemical are toxicity assays, which measure the effect of the chemical under specified test conditions. The toxicity of a chemical is proportional to the severity of the chemical on the monitored response of the test organism(s). Toxicity assays utilize test species that include rats, fish, invertebrates, microbes, and seeds. The assays may use single or multiple species of test organisms. Although no single bioassay procedure can provide a comprehensive toxicity evaluation of a soil or sludge, a valid toxicity screening test can provide information about the relative toxicity of a compound, can help predict noninhibitory chemical application rates, and can help estimate the feasibility of bioremediation for a specific situation.

Toxicity assays using bacteria as the test organism are rapid, easy to use, cost-effective, and use a statistically significant number of test organisms. One such bacterial toxicity assay method is the Microtox© assay. This method is relatively simple, rapid, and inexpensive. The use of the Microtox© procedure [4] to screen and predict the treatability potential of waste in soil and of contaminated soil has been evaluated and found to be satisfactory [5,6].

The Microtox© system is a standardized toxicity test that utilizes marine luminescent bacteria (*Photobacterium phosphoreum*) as indicator organisms. Bioluminescence of this test organism depends on a complex chain of biochemical reactions involving the luciferin-luciferase system. Chemical inhibition of any of the involved biochemical reactions causes a reduction in bacterial luminescence. The Microtox© toxicity assessment considers the physiological effect of a toxicant and not just mortality.

This method utilizes an instrumental approach in which the indicator organisms are handled as chemical reagents. Suspensions of about one million bioluminescent organisms are "challenged" by addition of serial dilutions of an aqueous sample. A temperature controlled photometric device quantifies the light output in each suspension before and after sample addition. Reduction of light output reflects physiological inhibition, which indicates the presence of toxic constituents in the sample. Such tests do not provide information on toxicity from a human health or an environmental standpoint. Rather, they are used as a relative toxicity screening method for contaminated soil and sludges and to identify the relative toxicity reduction that occurs when chemicals and wastes are managed by bioremediation processes.

WATER

The presence of water is essential for microbial activity. At low water concentrations, nonspore-forming microorganisms will die or their concentrations will be reduced greatly. At high water concentrations, such as at or near saturation, the pores of the soil are filled with water and diffusion of oxygen from the atmosphere is restricted. Under the later situation, anaerobic rather than aerobic conditions will occur in soils unless oxygen is mechanically added. The water content of soil typically ranges from 15 to 35 volume percent. At 35%, most soils are water saturated. The water content can drop below 15% under arid conditions. Soil water content is commonly expressed as a percent of soil water-holding capacity. A soil water-holding capacity range of 25% to 100% is the equivalent to a range of about 7 to 28 volume percent.

Over the range of about 30% to 90% of the water-holding capacity of the soil, the moisture content has little effect on biodegradation rates [7]. In solid phase bioremediation systems, moisture control may be necessary to achieve optimum biodegradation rates.

OXYGEN

Adequate oxygen and aerobic conditions in bioremediation systems are important to: (1) avoid odors produced by anaerobic conditions and (2) produce the most oxidized end-products. Anaerobic conditions and degradation can occur in bioremediation systems but should be avoided since anaerobic biodegradation is slower and less complete, and under reduced conditions most metals are more water-soluble.

For solid phase bioremediation systems used for contaminated soils, aerobic conditions are maintained by: (1) mechanically mixing the material, such as in composting or tilling the mixture, as in land treatment, (2) avoiding saturating the mixture of water, and (3) maintaining the quantity of degradable material in the mixture such that the oxygen demand does not exceed the rate at which oxygen is transferred from the atmosphere to the mixture.

For slurry phase bioremediation systems used for sludges, oxygen is added by mechanical mixers and aeration systems. These systems also mix the liquid slurry to keep the particles in suspension and enhance oxygen transfer. Oxygen transfer in slurry systems usually is less than in dilute or clean water systems and the aeration equipment should be sized to transfer the necessary amount of oxygen under operating and not clean water conditions. To avoid oxygen-limiting conditions, the dissolved oxygen concentrations in slurry reactors should be maintained above about 0.5 mg/L.

ELECTRON ACCEPTORS

In a bioremediation system, microorganisms metabolize organic compounds to obtain biological energy for microbial growth and maintenance. In this process, electrons from incompletely oxidized (reduced) compounds are transferred along respiratory electron transport chains, energy is captured by the microorganisms, and oxidized end-products, such as carbon dioxide (CO_2) result. Under aerobic conditions, oxygen (O_2) is the terminal electron acceptor. When O_2 is not available, nitrate (NO_3^-), iron (Fe^{3+}), manganese (Mn^{2+}), and sulfate (SO_4^{2-}) can act as electron acceptors if the organisms have the appropriate enzyme systems.

Microbial degradation removes oxygen from media in the bioremediation processes. If oxygen is not replenished, the systems become devoid of oxygen, reduced conditions result, and other substances are used as terminal electron acceptors. To maintain aerobic conditions, oxygen must be added to the bioremediation systems by the methods noted in the previous section.

The loss of oxygen from the system also can cause a change in the microbial population. Facultative bacteria, which can use electron acceptors such as nitrate, sulfate, or oxygen, and anaerobic organisms become the dominant microbial populations. In bioremediation systems, the vast majority of the microorganisms are facultative.

pH

The optimum pH for microbial degradation is around neutral, generally in the range of 6 to 8. Biodegradation can occur outside this range although at reduced rates. In bioremediation systems, pH control rarely is needed unless very acid or alkaline conditions are encountered. pH control can be needed if the biological activity causes a marked change in the pH.

NUTRIENTS

Microbial metabolism and growth require adequate macro- and micro-nutrients. The soil normally supplies adequate micronutrients such as trace metals and minerals. However, it cannot be assumed that all soils and sludges have adequate macronutrients such as nitrogen (N), phosphorus (P), or potassium (K). Organics in contaminated soils and sludges can be high in carbonaceous content but low in N and P.

The need for additional N and P is controlled by: (1) the amount of N and P in the contaminated soil and sludges being treated and the rate that they are available and (2) the amount and rate at which organic carbon is degraded in the mixture. By estimating or measuring these factors, the need for additional N and P can be determined. A carbon to nitrogen (C/N) ratio of greater than 35–40 generally indicates inadequate nitrogen. A C/N/P ratio of about 100:10:1 will provide adequate nutrients in a bioremediation system. If inadequate nitrogen exists, the deficiency can be remedied by adding a chemical fertilizer such as ammonium sulfate or ammonium diphosphate.

TEMPERATURE

Biodegradation rates are affected by temperature, generally changing by a factor of two for a change of 10°C in the temperature range of 5°C to 30°C. Low temperature is not lethal to microorganisms but will drastically reduce biodegradation rates. Biodegradation is essentially zero at freezing temperatures. Insulated reactors can be used to keep the reactor temperature above external air temperatures.

In bioremediation systems, temperature control rarely is practiced. Different systems will operate at different temperatures, however. For example, solid phase composting type systems may operate at temperatures that are above ambient. The temperature in bioremediation systems normally is near the ambient temperature and will change slowly as ambient temperatures change. Abrupt temperature changes are unlikely.

WATER SOLUBILITY

Microbial degradation is carried out by microbial enzymes. In bioremediation systems, the substances to be degraded must contact or be transported to and into the microbial cell. Thus, the water solubility of the chemical is important since only the water-soluble fraction of a chemical is readily degradable.

Many of the chemicals in contaminated soils and sludges are not very water soluble. This low solubility can reduce the availability of the chemical for degradation and will be a controlling parameter in bioremediation. The aqueous solubilities of chemicals commonly found in contaminated soils and sludges are noted in Table 4.5. The aqueous solubility decreases as the chemical becomes more complex.

Chemicals that have low water solubility generally are soluble in organic solvents. A commonly used measure of the relative organic solvent solubility is the octanol-water partition coefficient, K_{ow}, which is defined as the ratio between the concentrations of the compound in equal parts of octanol and water. A value greater than 1 indicates that the compound is more readily soluble in octanol than in water. Octanol models the behavior of other hydrocarbons and is used as a model nonpolar organic hydrocarbon solvent. K_{ow} values for many chemicals also are noted in Table 4.5.

The octanol-water partition coefficient is a key parameter in studies of the environmental fate of chemicals. K_{ow} has been found to be related to water solubility, soil and sediment adsorption coefficients, and bioconcentration factors (BCF) [8].

Chemicals with low K_{ow} values, i.e., less than about 10, have high water solubilities, small adsorption coefficients, and small BCF values. Chemicals with high K_{ow} values, i.e., greater than about 1000, are very hydrophobic with low water solubilities and high sorption coefficients.

The aqueous solubility (S) of a chemical can be estimated from K_{ow} values. A number of equations have been developed to correlate these two parameters. The equations generally take the form of:

$$\log S = (a) \log K_{ow} + b \qquad (4.1)$$

where a and b are empirical regression equation constants.

SORPTION

Sorption is another parameter that affects organic chemical degradation. This occurs because the greater the extent to which a chemical is sorbed

TABLE 4.5. Solubility and Octanol-Water Partition Coefficients of Organic Chemicals.*

Compound	Aqueous Solubility (mg/L)	K_{ow}
Monocyclic Aromatics		
Benzene	1,787	135
Toluene	515	540
o-Xylene	213	1,320
p-Xylene	185	1,410
m-Xylene	146	1,585
Ethylbenzene	110	1,410
Other Halogenated Compounds		
Trichloroethylene	1,100	260
Pentachlorophenol	14	132,000
Polyaromatic Hydrocarbons (PAH)		
Naphthalene	31	2,000
Acenaphthylene	3.93	5,500
Acenaphthene	3.42	8,300
Fluorene	1.98	15,100
Phenanthrene	1.29	28,800
Fluoranthene	0.26	79,400
Pyrene	0.135	75,900
Anthracene	0.066	28,200
Benz[a]anthracene	0.014	407,000
Benzo[a]pyrene	0.0038	933,000
Chrysene	0.002	407,000
Benzo[b]fluoranthene	0.001	3,720,000
Benzo[k]fluoranthene	0.0003	6,920,00
Benzo[ghi]perylene	0.00026	17,000,000
Benz[a]anthracene	0.014	407,000

*Adapted from Tetra Tech (1989) [9].

to a soil, the longer it is retained in soil and the greater the amount of time that is available for degradation. Thus even chemicals that have slow degradation rates (long half-lives) can undergo satisfactory degradation in contaminated soil bioremediation systems.

The mobility of a chemical in a soil system can be expressed in terms of a retardation factor, which can be calculated as:

$$R = 1 + \frac{K_p \cdot \varrho_b}{\Theta} \qquad (4.2)$$

where R = retardation factor (unitless), K_p = soil-water partition coefficient (ml/g), ϱ_b = dry bulk density of permeable material (g/ml), and Θ = volumetric moisture content (decimal fraction). The partition coefficient, K_p, also is known as the sorption coefficient of a chemical to soil.

If a chemical does not interact with the soil, K_p equals 0 and the retardation factor equals 1. Examples of such chemicals are chlorides and nitrates. Chemicals with high sorption coefficients are more tightly bound to the soil, are less mobile, and have lower degradation rates.

Adsorption is a surface phenomenon in which matter is extracted from one phase and concentrated at the surface of the second. Adsorption of a chemical from solution onto a solid occurs as the result of the lyophobic (solvent-disliking) character of the chemical relative to the particular solvent, or of a high affinity of the chemical for the solid.

Adsorption is modeled by an isotherm, which is a means of describing changes in adsorption at constant temperature. Adsorption of many chemicals found in contaminated soils commonly can be described by a Freundlick isotherm

$$q = KC^{1/n} \tag{4.3}$$

where q = mass of chemical sorbed per mass at equilibrium, K = adsorption coefficient, C = concentration of chemical in the liquid, and n = an empirical constant representing adsorption intensity.

The adsorption coefficient (K) can be normalized to the organic carbon content of the mixture (K_{oc}):

$$K = (K_{oc}) (f_{oc}) \tag{4.4}$$

where f_{oc} is the fraction of organic carbon content of the soil. This relationship works well for soils with organic carbon contents above about 0.5%.

Thus, it is possible to express the tendency of a chemical to be adsorbed in terms of K_{oc}, which is largely independent of the properties of the soil. K_{oc} may be thought of as the ratio of: (1) the amount of chemical adsorbed per unit weight of organic carbon (oc) in the soil to (2) the concentration of the chemical in solution at equilibrium:

$$K_{oc} = \frac{\text{mg adsorbed/g organic carbon}}{\text{mg/mL solution}} \tag{4.5}$$

K_{oc} is a constant only if the adsorption isotherms are linear. Generally, linearity occurs when the equilibrium aqueous phase organic chemical concentration is below one half of the pure compound aqueous solubility of the chemical.

K_{oc} also has been related to the aqueous solubility (S) and the K_{ow} of a chemical. The relationships generally take the form of:

$$\log K_{oc} = (c) \log (S \text{ or } K_{ow}) + d \tag{4.6}$$

where c and d are empirical regression constants.

VOLATILIZATION

Volatile constituents can be present in many soils and sludges. Such constituents can be released to the atmosphere during bioremediation processes. Controls on volatile losses may be needed as part of bioremediation systems to meet regulatory requirements.

The occurrence of volatilization in such systems is a function of the vapor pressure of the chemical and the contact between the chemical in the soil and the gaseous phase or atmosphere. Bioremediation systems are mixed and/or aerated to provide adequate oxygen input and aerobic conditions. As a result, such systems offer excellent opportunities for volatile compounds to be released.

CHEMICAL LOSS RATES

Results from field systems and laboratory and pilot plant bioremediation systems indicate that the losses of organic compounds can be described by first-order rate reactions in which the rate of loss of a chemical is proportional to the chemical concentration:

$$dc/dt = -kc \qquad (4.7)$$

where c = chemical concentration (mass/mass), t = time, and k = first-order rate constant (time^{-1}).

Chemical loss rates usually are discussed in terms of half-life ($t_{1/2}$), i.e., the time required to degrade or lose one-half of the chemical concentration. Mathematically, half-life can be calculated from the first-order rate constant:

$$t_{1/2} = 0.693/k \qquad (4.8)$$

Examples of half-life values of chemicals in contaminated soil bioremediation systems are indicated in Table 4.6.

Most chemical loss rates that are reported represent total chemical loss, and rarely are specific loss mechanisms identified. The losses can be due to biodegradation, chemical degradation, hydrolysis, photolysis, and volatilization. While it commonly is assumed that microbial degradation is the major loss mechanism, volatilization also can be an important loss mechanism for chemicals with a high vapor pressure.

Since remediation cleanup goals usually are related to concentration criteria, total loss rates are useful in determining the feasibility of bioremediation processes.

TABLE 4.6. Chemical Loss Half-Life Values in Contaminated Soil Bioremediation Systems.*

Organic	Half-Life Values ($t_{1/2}$) (days)
Benzene	0.1–1.0
Ethylbenzene	6.1
Toluene	6.4
o-Xylene	93.9
m,p-Xylene	8.5–14.7
Anthracene	9–53
Benzo(a)anthracene	41–231
Chrysene	5.5–116
1-Methylnaphthalene	12.6
Naphthalene	8–30
Phenanthrene	23–69
Pyrene	10–100

*Data from situations where relatively fresh refinery wastes were being land treated and evaluated.

SUMMARY

Bioremediation processes can successfully treat contaminated soils and sludges. The microorganisms in such systems will grow and metabolize the organics present if environmental conditions are suitable. Aerobic conditions are desirable and are accomplished by providing mixing and mechanical aeration.

Nutrients, such as nitrogen and phosphorus, may be needed. Toxic conditions should not exist, and pH and temperatures should be controlled as needed.

Many organics in contaminated soils and sludges have low water solubility which can limit degradation and are tightly sorbed to the organic matter. The sorption constant and water solubilities of chemicals can be estimated from known partition coefficients such as K_{ow} and K_{oc}.

Loss rates, rather than biodegradation rates, are commonly reported. Such loss rates can be used to determine whether performance goals can be achieved and the overall feasibility of a bioremediation process.

With an understanding of how the fundamentals described in the above sections can relate to and affect the performance of bioremediation processes, such processes can be technically and economically feasible.

BIOREMEDIATION PROCESS SELECTION

GENERAL APPROACH

A logic format can be used to identify the bioremediation technology most appropriate for site-specific conditions. The procedure considers the type of contaminant, subsurface conditions, groundwater conditions, types of development on and adjacent to the site, permit requirements, and relative costs. In addition, the procedure helps indicate: (1) additional information that may be needed for a final decision and (2) the type of detailed investigations and evaluations that will obtain needed additional information. In most cases, the final selection of and design for an appropriate bioremediation process will depend on the results of more detailed investigations.

Figure 4.1 indicates the major steps for determining the most appropriate bioremediation technology for a site. The steps involve evaluating: chemical composition of the soil or sludges, whether the constituents are

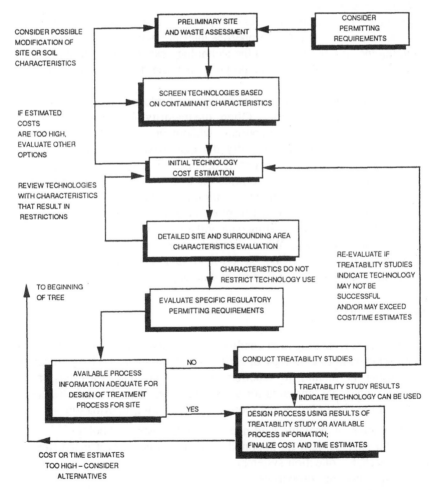

Figure 4.1 Schematic of steps in selecting appropriate bioremediation process for a site.

considered hazardous or nonhazardous, and physical handling characteristics. Site information will identify the choice of in-situ or on-site treatment technologies and the engineering considerations for such a technology.

SOIL AND SLUDGE CHARACTERISTICS

The chemical and physical characteristics of soils and sludges affect handling and treatment requirements. When making an initial evaluation of applicable bioremediation processes, sufficient information on the characteristics of the soil or sludges must be available to proceed with the technology evaluations. If adequate characteristic information is not available, a sampling program should be initiated before proceeding with the technology evaluations.

Soil and sludge characteristics can adversely affect process performance in one of three ways: (1) the properties interfere with the processes used in treating the soil or sludge; (2) the technology cannot adequately treat/immobilize a chemical; or (3) the property increases the time and/or costs of the process.

The first step is to characterize background information for factors that are important to the selection. Often, the majority of this information can be obtained from available sources, without the need for an immediate detailed site investigation. Usually the most useful data concerning the type and extent of contamination is obtained directly from the initial investigations undertaken to determine the extent to which a contamination problem exists. Much of the information required for technology screening often can be approximated from existing data bases and site observations. Utilization of the existing data will always be the most cost-effective first investigation step. When detailed investigations are required, the available data will provide background to assure that the efforts are directed towards the most important data gaps.

SITE CONDITIONS

Site and surrounding area conditions can severely restrict the choice of a bioremediation technology or series of technologies by: (1) significantly increasing the monitoring requirements or the pretreatment needs or (2) having inadequate land area for a specific technology. The initial assessment of the site to be treated requires the following information:

- amount of available land needed to implement technology
- for sludges, how an impoundment was initially constructed (liners, leachate collection systems)
- soil type and properties

- depth to seasonally high water table
- location of site in relation to inhabited areas
- distance from nearest drinking water source
- regional climatic data

The amount of available land influences the ability to place tanks and equipment and affects the choice of in-situ versus on-site treatment. The requirements for a liner beneath a constructed treatment unit may influence subsequent decisions about performing treatment in a reactor and the choice of equipment for treatment. Soil type and local hydrogeologic characteristics will affect the final choice of treatment technology and the type of engineering considerations for its implementation. The presence of a seasonally high water table may negate the possibility of in-situ treatment. Climatic data is important when considering bioremediation methods exposed to ambient conditions. The amount of rainfall, the length of the rainy season, the expected maximum and minimum temperatures, and the length of the warm and cold months will influence the design, implementation, and operation of any bioremediation system.

The decision to implement a particular treatment option also is affected by regional development and proximity to residential areas. The decision for in-situ versus on-site treatment is strongly influenced by the physical factors of the site, the soil characteristics, and the applicable regulatory requirements. Additionally, local opinion can play a significant role in the acceptance of a technology. Consideration of historical public reactions to proposed treatment processes should be considered early in the decision process.

REGULATORY REQUIREMENTS

One of the first steps in selecting an appropriate technology for a site is to identify the existing state and federal permitting needs and requirements. There may be specific data and types of evaluations that will be required to obtain a permit for a site. After an appropriate technology has been identified (or a treatment train determined), the specific regulatory requirement for a site and process must be reevaluated. It is likely that permits will be needed: (1) for the soil or sludge bioremediation process; (2) to assure protection of surface and groundwater; and (3) to avoid air quality impact.

The choice of a treatment technology and its implementation will be affected by the regulatory requirements for cleanup and closure. Hazardous sludges must be treated to specific Constituent Action Levels (CALs) or meet Best Demonstrated Available Technology (BDAT) standards for subsequent land disposal with a selected process. The range of treatment possibilities for a nonhazardous waste will be greater than for a hazardous waste. Land disposal of the treated residues for a nonhazardous waste will be possible at any non-RCRA land disposal facility. Land disposal in RCRA permitted facilities also will be possible for those hazardous wastes that can be treated to meet the BDAT standards. Figure 4.2 indicates the final treatment and disposal options for a bioremediation process treating a hazardous waste contaminated slurry. If the residues can be delisted based on health-based criteria, the residues can go to a nonhazardous waste land disposal facility. These options are noted in Figure 4.3.

The decision of in-situ versus on-site treatment affects the final treatment and closure options for contaminated soils and sludges. A soil that is not moved from its "natural" location during treatment and ultimate disposal will not be subject to the land disposal restrictions and the minimal technology requirements (MTR) for closure. MTRs require that certain minimum construction practices and operations be conducted at a landfill or impoundment to ensure protection of human health and the environment. These minimum technology requirements include: (1) double liner system; (2) leachate collection system; and (3) groundwater monitoring and closure requirements.

By not moving the soils from the original location, no placement or generation of hazardous waste will occur, and the site can be closed as a solid waste landfill. This will require the MTRs for landfill capping and post-closure groundwater monitoring.

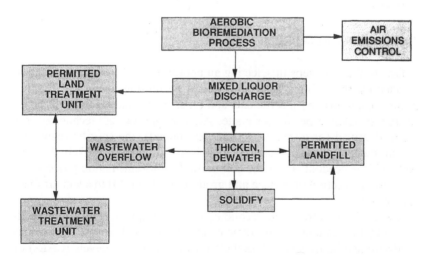

Figure 4.2 Treatment and disposal options using an aerobic bioremediation process.

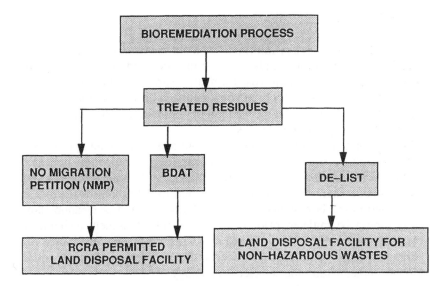

Figure 4.3 Disposal options for residues from the bioremediation of hazardous wastes.

TREATABILITY STUDIES

Important in the choice of a technology are the results of treatability studies. A treatability study should be considered, and if needed, undertaken early in the decision-making process. If adequate process information is available from personal experience or from the literature, and the soil or sludge is adequately characterized, a treatability study may not be necessary to proceed with the design and cost estimating phase. If sufficient soil or sludge characterization data are not available, a treatability study is very important. The results from the study can provide data on process design and operating parameters, materials handling, process optimization, and the time and conditions required to reach specific treatment goals.

Treatability studies are laboratory, bench, or pilot scale studies using a representative sample of the soil or sludge to be treated, and the various parameters associated with each technology. Small biological reactors are used to determine the estimated time required for degradation and to evaluate the need for pH adjustment and nutrient addition. Air emissions control needs can be evaluated at this time. The parameters evaluated during the treatability study will depend upon the specific technologies being considered.

FINAL FACTORS

Once all of the necessary information on the soil or sludge characterization, from the treatability studies and on the site conditions, is available, design and situation specific cost estimation can be completed. This includes determining the appropriate treatment and/or disposal for the residuals, which is dependent on the technology selected, and the available on-site treatment/disposal alternatives.

At this point, the technical and economic feasibility of an appropriate bioremediation process should be apparent. If this evaluation indicates that such a process is feasible, a more detailed investigation of the site and soil or sludge characteristics may be desired to determine final process design and costs. For regulated wastes, the negotiation of cleanup standards should be initiated prior to making a final treatment technology selection. Information about cleanup standards, costs, and operational requirements also will influence the permitting and design decisions made prior to implementing the selected technology.

BIOREMEDIATION PROCESS DESCRIPTIONS

As noted earlier, bioremediation processes can be categorized as solid and slurry phase processes. The solid phase processes include: (1) surface soil and on-site modifications of land treatment and (2) a modification of the basic composting process. Slurry phase processes include liquid-solids systems that can be used in existing impoundments or in tanks. The following describes each process and its potential applications.

SOLID PHASE LAND TREATMENT

Land treatment is a managed technology that involves controlled application of a waste on the soil surface and/or the incorporation of the waste or contaminated soil into the upper soil zone. It is not the indiscriminate dumping of waste on land, and it is not landfilling. Land treatment technology relies on the dynamic physical, chemical, and biological processes occurring in the soil. As a result, the constituents in the applied wastes are degraded, immobilized, or transformed to environmentally acceptable components.

The design and operation of a land treatment facility is based on sound scientific and engineering principles as well as on extensive practical field experience. A land treatment site is designed and operated: (1) to maximize waste degradation and immobilization, (2) to minimize release of dust and volatile compounds as well as percolation of water-soluble waste compounds, and (3) to control surface water runoff. Land treatment is a bioremediation process that can be a viable management practice for the treatment and disposal of hazardous and nonhazardous wastes. Land treatment has been successfully practiced in all major climatic regions of the United States, Europe, and Canada under a wide range of hydrogeologic conditions. In the U.S., approximately 200 industrial land treatment systems are in operation, including over 100 that have been at petroleum refining facilities.

Both surface soil and on-site land treatment processes are among the more widely used bioremediation technologies. *Surface soil land treatment* bioremediation involves: (1) keeping contaminated surface soils in place, (2) if needed, adding nutrients to assure adequate biodegradation and adjusting the pH toward neutral conditions, (3) tilling the soil periodically to increase the availability of oxygen and nutrients to the soil microorganisms, and (4) possibly irrigating to assure adequate moisture for microbial degradation. The organisms involved in the degradation are indigenous organisms unless none that can degrade the organics exist in the soil due to prior toxicity. In such cases, the toxicity needs to be reduced by: (1) adding uncontaminated soil or (2) adjusting the environmental conditions and adding acclimated organisms. The rate and extent to which the organics in the soil are lost will depend on the combined effect of volatilization, sorption, and biological degradation.

Both degradation and immobilization occur in a land treatment unit with most of the residual organics and metals remaining in the upper layers of the soil, essentially in or near the zone mixed by tilling. Figure 4.4 illustrates typical patterns of organics and metals in the soil from a land treatment unit that had treated petroleum refining wastes and residues for several decades.

Land treatment units also minimize the leaching of constituents applied to the unit or in the contaminated soil being bioremediated. Figure 4.5 illustrates the poor mobility of metals in soils, as measured by the toxicity characteristic leaching procedure (TCLP), at various depths at an industry land treatment unit.

Site runoff must be contained and managed at all waste treatment operations. The runoff water that is collected may be allowed to evaporate on site, discharged to a wastewater treatment system, or reapplied to the land treatment site.

Surface soil land treatment units do not have a liner; however *on-site land treatment* units do. An on-site land treatment unit is a constructed unit that contains contaminated soil or clean soil and wastes being bioremediated and has walls, a drainage system, and a leachate collection

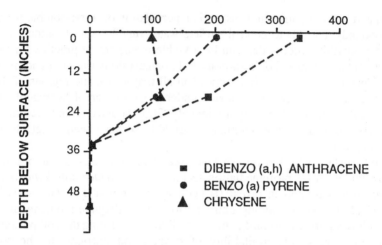

Figure 4.4a Average soil PAH concentrations as a function of depth – data from an active hazardous waste land treatment site [10].

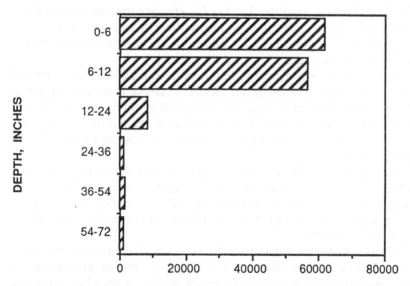

Figure 4.4b Oil and grease (mg/kg) as a function of depth – soil cores from an active hazardous waste land treatment site [10].

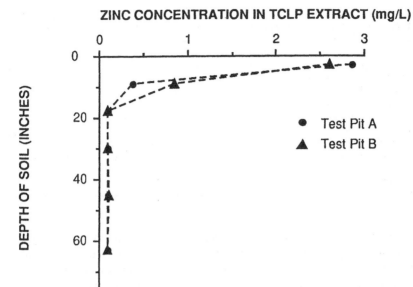

Figure 4.5 Zinc concentrations in the TCLP extracts from soil cores at a hazardous waste land treatment site [10].

system. It generally is above ground and allows more complete control of the process. An on-site land treatment unit also has been called a prepared bed bioremediation process.

An on-site land treatment unit usually is constructed adjacent to the site requiring bioremediation to minimize transportation costs and to provide better technical and managerial control of the process. Except for the fact that the on-site unit is constructed above ground, the fundamentals and operations of the on-site unit are the same as for the surface soil unit.

Both surface soil and on-site land treatment systems have been used successfully to bioremediate: (1) contaminated soils at spill sites, (2) industrial wastes and residues, and (3) soils at surface impoundments and lagoons that are being closed. Land treatment has been an effective bioremediation process for wood preserving waste contaminated soil and sludge. The loss rates ($t_{1/2}$) of the PAH compounds in land treatment plots remediating such soil and sludge are summarized in Table 4.7. The results are indicative of the type of losses that occur under field situations.

COMPOSTING

Composting as a bioremediation process for contaminated soils and industrial wastes and residues is similar to the process used for composting

TABLE 4.7. Loss Rates of PAH Compounds from the Bioremediation of Wood Preserving Waste Contaminated Soil and Sludges Using Land Treatment.

Compound	Half-Life ($t_{1/2}$, days)	Compound	Half-Life ($t_{1/2}$, days)
Naphthalene	95	Chrysene	95
Acenaphthylene	105	Benzo(b)fluoranthene	290
Acenaphthene	40	Benzo(k)fluoranthene	140
Phenanthrene	25	Benzo(a)pyrene	100
Anthracene	35	Dibenzo(a,h)anthracene	260
Fluoranthene	40	Benzo(g,h,i)perylene	360
Pyrene	40	Ideno(1,2,3-cd)pyrene	600
Benzo(a)anthracene	50		

of leaves, garbage, and food processing residues. The main differences are that high temperatures rarely are achieved in contaminated soil biomediation composting and the purpose is degradation and loss of specific organic compounds rather than stabilization to produce a mulch or soil conditioner. At the completion of a bioremediation composting process, the treated material must be disposed of in an environmentally sound manner. However, both degradation and immobilization have occurred in this bioremediation process and the composted material should not cause surface or groundwater problems at its ultimate disposal site.

The essential elements of composting are the same as for any bioremediation process: (1) moisture, (2) aeration, (3) acclimated organisms, (4) satisfactory carbon-nitrogen-phosphorus balance, and (5) nontoxic conditions. The characteristics of contaminated soils or residues considered suitable for composting are: (1) constituents able to be lost by volatilization or degradation, or immobilized in the system, (2) a low amount of free liquid so that aerobic conditions can be maintained, (3) a high ratio of inert solids to biodegradable organics, and (4) a mixture that can be easily broken up by mechanical turning and/or is porous to allow air to move through the composting solids.

Typical composting systems that can be used for bioremediation are the windrow, the Beltsville, and the in-vessel systems. The windrow system is an open system, with periodic turning of the compost mix pile and no forced air. The Beltsville system is an open pile with an air distribution system under the pile. Air is sucked through the pile from the atmosphere and exhausted through a blower generally to an air pollution control system. The in-vessel system occurs in a closed or open vessel in which mechanical mixing may occur and/or air is forced through the mixture by blowers. Bulking agents commonly are added to increase the porosity and assist the flow of air to maintain aerobic conditions.

For bioremediation composting systems, adequate operational control as well as control of all emissions, such as leachate and off-gases, is desirable. In-vessel systems provide such control. If gaseous emissions are not a concern, then a windrow or Beltsville system with positive leachate control can be satisfactory.

Bioremediation composting systems commonly are constructed and operated on-site so as to minimize transportation costs. The systems can be operated as continuous or batch units. Contaminated soil or residues are excavated from the site, or wastes and residues are added periodically for treatment and remediation. After meeting specified cleanup criteria, the remediated mixture may be returned to the site or disposed of in a landfill or other acceptable location.

Composting has been used to remediate soils contaminated with diesel fuel and similar petroleum products.

SLURRY PHASE PROCESSES

Liquid-solids treatment systems are slurry phase bioremediation systems operated to maximize mass transfer rates and contact between contaminants and microorganisms capable of degrading the contaminants. Because solids can be treated rapidly in contained reactors, much less area is needed for this on-site remediation process than for land treatment. An advantage of liquid-solids systems using tanks is that regulatory concerns related to the land disposal of hazardous wastes or to the contamination of groundwater may be eliminated.

Liquid-solids contact treatment is analogous to conventional biological suspended growth treatment (e.g., activated sludge). These units are designed to relieve the factors commonly limiting microbial growth and activity in soil, principally, the availability of carbon sources, inorganic nutrients, and oxygen. To achieve this goal, the sludges or contaminated soils are suspended in a slurry form and mixed. The mixing and aeration also prevent oxygen transfer limitations. Mixing can be provided by aeration alone or by aeration and mechanical mixing. Aeration is provided by floating or submerged aerators or by compressors and sparges. Chemicals added to liquid slurry reactors can include nutrients and neutralizing agents to relieve limitations to microbial activity.

Liquid-solids systems are a relatively new approach for the remediation of contaminated organic materials. There have been several applications of these systems for the treatment of wood-treating, coal tar, and petroleum wastes. The most frequently used process configuration is a simple batch above-ground tank or aeration of an existing pit or lagoon.

After the bioremediation cleanup goals are achieved in a batch reactor, the aeration and mixing are discontinued and the solids are allowed to settle. The treated liquid is decanted and discharged while the treated solids can be further treated by conventional land treatment or disposed of on-site. If the process is used in a lagoon or surface impoundment, it may be possible to decant the liquid and leave the treated solids in place.

Liquid-solids systems are suitable for contaminated slurries, such as may be in pits, ponds, and lagoons, and some contaminated soils. The technology will treat polyaromatic hydrocarbons, naphthalene, phenols, benzene, toluene, xylene, and ethylbenzene. The degree of treatment of any one chemical is a direct function of its solubility in water and its rate of biodegradation.

Because of the mixing and aeration in these systems, volatilization is a significant factor in the removal of organics. Monitoring and management of the gaseous emissions may be necessary at locations where air quality concerns exist.

The phenomena that govern the process are mass transfer of organics from the solid phase to the aqueous phase and biodegradation of the aqueous phase organics. The design considerations important to properly apply these systems are:

- the physical characteristics of the liquid-solids feed, particularly the fraction of the mix that is organic, the distribution of organics within the mixed slurry, and the viscosity and surface tension of the hydrocarbon phase
- the nutrient, temperature, and oxygen requirements to achieve the optimal biodegradation
- the solids and hydraulic residence times to achieve adequate treatment and meet the cleanup goals
- the characteristics of the offgas
- the degree of mixing (energy) required to keep the solids in suspension

A liquid-solids reactor is one part of a slurry phase treatment system. Consideration must be given to how the soil or sludge will be removed from the source and transmitted to the reactor, if pretreatment such as thickening is desirable, the emission controls that may be needed, and what steps may be needed to handle and dispose of the treated slurry. The process options are indicated in Table 4.8 and the disposal options for the treated residues were shown in Figure 4.2.

SUMMARY

There are several bioremediation systems that can be used for contaminated soils and slurries. The benefits of such systems are: (1) degrada-

TABLE 4.8. Soil Slurry Treatment Process Options.

Process Step	Process Options
Excavation and Transport	Dredge, earth moving equipment
Pretreatment	Thicken or dilute
Emission Controls	Carbon adsorption, biofiltration, combustion, recirculation of off-gas to slurry reactor
Slurry Treatment	Batch, semicontinuous, continuous
Post-Treatment	Land treatment, coagulate/flocculate, thicken, filter press, solidify, landfill

tion and immobilization of organics, (2) toxicity reduction, (3) source control and prevention of groundwater pollution, and (4) risk reduction.

Both solid phase and slurry phase processes are available and are being used with such materials. These processes can be used successfully if the fundamentals are understood and incorporated into design and operation. The costs of these systems generally are lower than alternative remediation methods. Because of the above benefits and the lower costs, bioremediation systems are an important technology for remediation of specific sites.

REFERENCES

1 U.S. Environmental Protection Agency, Technology Innovation Office. 1990. "Selected Data on Innovative Treatment Technologies for Superfund Source Control and Ground Water Remediation," Washington, D.C. (August).

2 Huddleston, R. L., C. A. Bleckmann and J. R. Wolfe. 1986. "Land Treatment Biological Degradation Processes," in: R. C. Loehr and J. F. Malina, Jr. (eds.) *Land Treatment: A Hazardous Waste Management Alternative, Water Resources Symposium No. 13*, Center for Research in Water Resources, The University of Texas at Austin, Austin, Texas.

3 Sims, J. L., R. C. Sims and J. E. Matthews. 1989. *Bioremediation of Contaminated Surface Soils*, Robert S. Kerr Environmental Research Laboratory, U.S. Environmental Protection Agency, EPA/600/9-89/073, Ada, Oklahoma.

4 Beckman Instruments, Inc. 1982. *Microtox© System Operating Manual*, Beckman Instruments, Inc., Carlsbad, California, 52 pages.

5 Matthews, J. E. and A. A. Bulich. 1984. "A Toxicity Reduction Test System to Assist in Predicting Land Treatability of Hazardous Organic Wastes," in: J. K. Petros, Jr. et al., (eds.) *Hazardous and Industrial Solid Waste Testing: Fourth Symposium*, Philadelphia, ASTM/STP 886.

6 Matthews, J. E. and L. Hasting. 1987. "Evaluation of Toxicity Test Procedure for Screening Treatability Potential of Waste in Soil," *Toxicity Assessment*, 2:265-281.

7 Dibble, J. T. and R. Bartha. 1979. "Effects of Environmental Parameters on the Biodegration of Oil Sludge," *App. and Environ. Microbiol.*, 37:729–739.

8 Lyman, W. J., W. F. Reehl and D. H. Rosenblatt. 1982. *Handbook of Chemical Property Estimation Methods*, McGraw-Hill Book Company.

9 Tetra Tech, Inc. 1989. "MYGRT Code Version 2.0: An IBM Code for Simulating Migration of Organic and Inorganic Chemicals in Groundwater," EPRI EN-6531, Final Report, Project 2879-2, Electric Power Research Institute, Palo Alto, CA.

10 Loehr, R. C., D. C. Erickson, L. A. Rogers and D. M. Kelmar. 1990. "Mobility and Degradation of Residues at Hazardous Waste Land Treatment Sites at Closure," Robert S. Kerr Environmental Research Laboratory, Ada, Oklahoma, PB 90-212-564AS, National Technical Information Service, Springfield, VA.

CHAPTER 5

Pump-and-Treat Systems for Groundwater Contaminants

INTRODUCTION

HYDROLOGIC, chemical, and biological fate and transport mechanisms must be evaluated at a waste site before any remedial action can be evaluated. Soil samples and cores should be collected and a group of boreholes and wells should be drilled and sampled to define the local hydrogeology and the source and extent of contaminant migration. Complete and accurate site investigation and characterization are of major significance for future remedial designs to be successful.

HYDROGEOLOGIC CHARACTERIZATION

The majority of information regarding the geologic formations and related surfaces underlying the site will be obtained through the description of sediment samples collected during drilling of soil borings and monitoring wells. It is worthwhile to describe all strata underlying the site to at least the maximum depth of known or potential contamination and generate a reliable and complete description of the subsurface geology. Continuous core samples can be collected using auger or rotary drilling methods.

The information obtained during the geologic investigation can be presented in vertical geologic cross sections and fence diagrams. Laboratory analysis of sediment or rock samples may include grain size analysis, hydraulic conductivity, plasticity, moisture content, dry density, clay mineralogy identification, and partition coefficients for pertinent chemicals. In addition to laboratory analysis, in situ analysis can also be made of the geology through borehole and other geophysical methods. These methods can provide many of the same parameters determined through laboratory analysis and can provide information on the extent of contaminant plumes, areas of buried trenching operations, and abandoned well locations.

One of the key elements affecting any remedial design is an accurate characterization of the groundwater flow system. This includes defining the physical parameters of the contaminated region such as hydraulic conductivity, storage coefficient, and aquifer thickness. Wells can be used for pumping tests and slug tests to determine hydraulic conductivity. Recharge rates and pumping rates are required along with any physical and hydraulic boundaries that affect the rate and direction of flow. Groundwater movement can be analyzed through the measurement of water levels in wells and piezometers. It is helpful to categorize wells according to the elevation and geologic formation of the screened interval so that the horizontal and vertical gradients of hydraulic potential can be analyzed separately. If there are enough measuring points, a contour map of the potentiometric surface of each aquifer can be prepared. The contour map can be evaluated to determine possible areas of groundwater recharge and discharge and to identify the direction of groundwater movement (Figure 5.1).

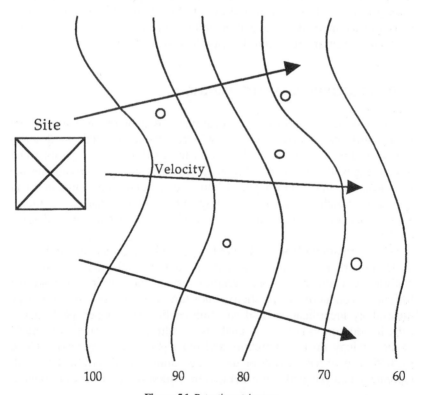

Figure 5.1 Potentiometric map.

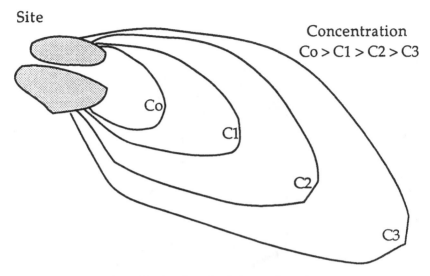

Figure 5.2 Plan view of typical contaminant plume.

The rate of groundwater flow is controlled by the porosity and hydraulic conductivity of the media through which it travels and by hydraulic gradients, which are influenced by recharge and discharge (Freeze and Cherry, 1979 [1] or Fetter, 1988 [2]). Because contaminants are advected with moving groundwater, it is important to characterize both the rate and direction of groundwater flow from areas of recharge (commonly via rainfall, surface pits or ponds, or irrigation) to areas of discharge (surface water or wells). Groundwater flows are greatly affected by subsurface heterogeneities, both vertical and horizontal, subsurface fractures, and other features that alter the average hydraulic conductivity of the aquifer. A typical plume of contamination is shown in Figure 5.2.

It is important to conduct a site characterization in a short time period since groundwater flow systems can vary with time. Seasonal variations in water levels as high as several feet can adversely impact a remediation scheme. For example, at a fuel spill site in Traverse City, Michigan, the reduction of the water table by three feet during the summer drought of 1988 caused a contaminated zone of residual fuel oil near the original water table to not be flushed by an injection well system (Figure 5.3).

SOURCE CHARACTERIZATION

A history of the contamination events should be prepared, to the extent possible, to define the types of waste and quantify their loadings to the

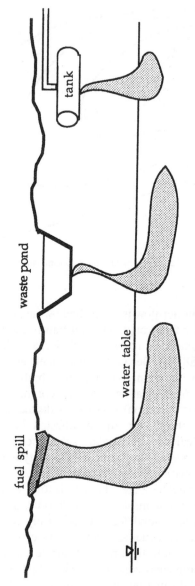

Figure 5.3 Contaminant sources.

86

system. It has become apparent in recent years that the source definition is one of the most important factors affecting groundwater contamination. It consists of defining the following: (1) the total mass or volume of chemicals released, (2) the total surface area affected by infiltration of chemicals, and (3) the time sequence of the release. Often, the release occurred so long ago that information is difficult to obtain since many sites have been abandoned or records were not kept. In these cases, one is forced to use hydrogeological data and plume contaminant data to compute what the release must have been based on a mass balance approach. In cases involving leaking tanks or pits, one can often estimate the volume of waste based on known pit or tank volumes and estimated infiltration rates. However, without detailed records, contaminant source definition is still the greatest source of error in designing a groundwater remedial scheme.

CONTAMINANT PLUME DATA

Information about the contaminant mix and spatial distribution of the plume is generally needed to select and analyze remedial alternatives during screening and detailed analysis phases. Physical and chemical properties of contaminants, such as density, solubility, and partitioning should be assessed because they influence plume movement. It should be recognized that some contaminants may not be detectable using routine analytical services, though they are present at levels that would be above cleanup levels, and special analytical services may have to be used.

Indicator chemicals are those site contaminants that are generally the most mobile and persistent; consequently, they reflect the likelihood of contamination at the site. Generally indicator chemicals are selected on the basis of mobility, persistence, ease of measurement, and volume of contaminants at the site. By initially identifying these constituents during the investigation, analytical costs can be reduced. During initial testing of the remedial action, however, samples should be analyzed for all contaminants present to ensure that indicator chemicals have been appropriately selected. Samples are generally analyzed once for total metals, cyanide, semi-volatiles, volatiles, and major anions and cations; periodically for those contaminants found at the site; and more frequently (e.g., during aquifer tests) for indicator chemicals. Before completing the remedial action, samples should be analyzed for all contaminants originally detected.

Quantitative characterization of the subsurface chemistry includes sampling both the vadose and saturated zones to determine the concentration distributions in groundwater, soil, and vadose water. A network of properly constructed monitoring wells needs to be installed to collect depth-discrete groundwater samples (U.S.EPA, 1986 [3]). Wells should be located in areas that will supply information on ambient (background) groundwater chemistry and on contamination in the plume. In the case of layered systems, it may be necessary to sample from wells properly completed in more than one vertical unit.

Soil and groundwater samples should be analyzed for the inorganic and organic parameters of concern from the waste stream. Of the various contaminants found in groundwater, the widely used industrial solvents and aromatic hydrocarbons from petroleum products are most common (Table 5.1). Chemical properties of the plume are necessary to characterize the transport mechanisms of the chemicals and to evaluate the feasibility of a remedial system. A number of properties influence the mobility of dissolved chemicals in groundwater and should be considered in detail for plume migration and cleanup. They include aqueous solubility, density, octanol-water partition coefficient, organic carbon partition coefficient, Henry's Law constant, and biodegradability. These are described in more detail in other chapters of the book.

To ensure proper quality assurance (QA) and quality control (QC) of groundwater samples, strict protocols must be followed in the field. The pH, temperature, and specific conductance of a sample should be mea-

TABLE 5.1. **Typical Organic Compounds Found in Waste Disposal Sites in the United States.**

Ground Water Contaminant
Thrichloroethene
Methylene chloride
Tetrachloroethene
Toluene
1,1-Dichloroethane
bis-(2-Ethylhexyl)phthalate
Benzene
1,2-trans-Dichloroethane
1,1,1-Trichloroethane
Chloroform
Ethyl benzene
1,2-Dichloroethane
1,1-Dichloroethane
Phenol
Vinyl chloride
Chlorobenzene
Di-n-butyl phthalate
Naphthalene
Chloroethane
Acetone
Xylene

sured. Ideally, before a sample is gathered, water should be extracted from the well until these parameters have stabilized. This will help ensure that the sample is from the formation. Proper sample storage and shipment to a qualified laboratory is also important. A sampling plan should address issues such as sampling frequency, locations, and statistical relevance of samples [4]. For more details on sampling guidance, see Cartwright and Shafer, 1987 [5]; Barcelona et al., 1983 [6]; and Barcelona et al., 1985 [7].

After analyzing the samples, the resulting concentration data should be mapped in two or three dimensions to determine the spatial distribution of contamination. These plume delineation maps and the results from aquifer slug and pump tests will yield estimates on plume migration and identify possible locations for injection or extraction wells to be used for remediation (Figure 5.2). The plume data may indicate that either adsorption or biodegradation is taking place based on the shape and relative concentration contours of various chemicals.

SOLUTE TRANSPORT PROCESSES

During the decade of the 1980s knowledge of the nature of organic contaminant transport in groundwater advanced considerably, due to more than a billion dollars spent on remedial site investigations at hundreds of sites. A few sites such as the Borden landfill, Cape Cod, the Denver arsenal, and Traverse City, Michigan, have been studied to such an extent that millions of dollars have been spent on site characterization, remedial investigations, and actual cleanups. These sites and others represent classic studies where the state of the science was significantly advanced while the site was being investigated and remediated (Mackay and Cherry, 1989 [8]).

Transport processes of major concern in groundwater were covered in Chapter 3 and include advection, dispersion, adsorption, biodegradation, and chemical reaction. The first two mechanisms have been analyzed in some detail for both laboratory and field conditions, while the latter three processes are the focus of current research efforts. The incorporation of these transport mechanisms into groundwater model formulations is described in more detail by Anderson, 1979 [9]; Anderson, 1984 [10]; Wang and Anderson, 1982 [11]; Bear, 1979 [12]; Freeze and Cherry, 1979 [1]; Bedient et al., 1985 [13]; and De Marsily, 1986 [14].

ADVECTION AND DISPERSION

Advection represents the movement of a contaminant within the bulk fluid according to the seepage velocity in the pore space. Figure 5.4 shows

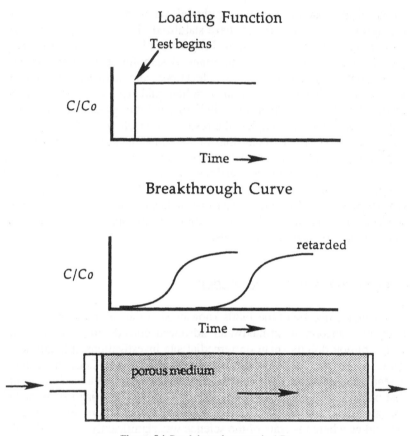

Figure 5.4 Breakthrough curves in 1-D.

the resulting breakthrough curve in one dimension (1-D) for solute transport in the absence of dispersion. There are certain cases in the field where an advective model provides a useful estimate of contaminant transport. Some models include the concept of arrival time by integration along known streamlines (Nelson, 1978 [15]). Others set up an induced flow field through injection or pumping and evaluate breakthrough curves by numerical integration along flow lines. Dispersion is not directly considered in these models, but results from the variation of velocity and arrival times in the flow field [16,17]. Capture zone approaches rely on concepts from well mechanics and do not consider dispersion since advection is considered to be dominant [18].

The dispersion process is described in detail in Bear, 1979 [12]; Anderson, 1979 [9]; Freeze and Cherry, 1979 [1]; Gelhar et al., 1979 [19]; Gelhar et al., 1985 [20]; Freyberg, 1986 [21]; and De Marsily, 1986 [14]. Dispersion is due mainly to heterogeneities in the medium that cause variations in flow velocities and flow paths. Laboratory column studies yield dispersivity estimates on the order of centimeters, while values in field studies may be on the order of meters. Figure 5.5 shows the typical spreading out

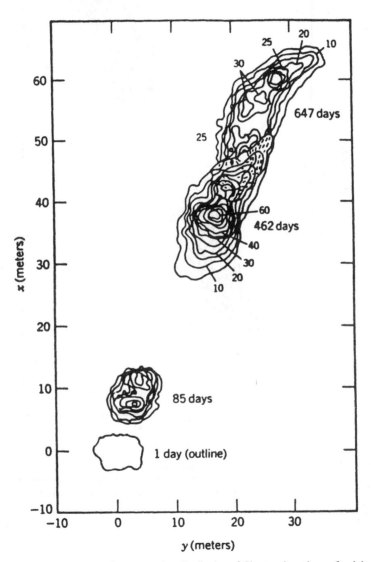

Figure 5.5 Vertically averaged concentration distribution of Cl⁻ at various times after injection and retardation of carbon tetrachloride (CTET) and tetrachloroethylene (PCE) (from Roberts and others, 1986, *Water Resources Res.*, v. 22, copyright by American Geophysical Union [26]).

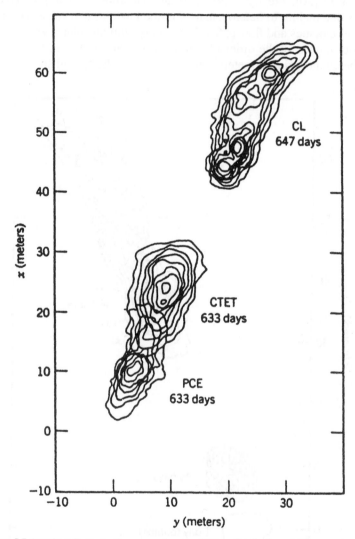

Figure 5.5 (continued) Vertically averaged concentration distribution of Cl⁻ at various times after injection and retardation of carbon tetrachloride (CTET) and tetrachloroethylene (PCE) (from Roberts and others, 1986, *Water Resources Res.*, v. 22, copyright by American Geophysical Union [26]).

of the contaminant front due to the dispersion of the contaminant in the porous media. Thus, the effect of dispersion is to cause the plume site to increase and the maximum concentration to decrease.

Dispersivity values have usually been set constant in transport models, but recent efforts by Gelhar et al., 1985 [20]; Dagan, 1984 [22]; Mackay et al., 1986 [23]; Freyberg, 1986 [21]; and Sudicky, 1986 [24] indicate that dispersivity depends on the distribution of heterogeneities and the scale of the field problem. Many investigators have worked on and contributed to the complex problem of estimating dispersivity from field tracer studies and pump tests, and both statistical and deterministic models have been postulated. One of the most extensive studies of the decade on dispersion effects was performed at the Borden landfill natural gradient test, and three-dimensional results were obtained using the method of moments for the sandy aquifer site. The asymptotic longitudinal dispersivity obtained by Freyberg, 1986 [21] from calibration to the Dagan, 1984 [22] model was 0.49 m.

ADSORPTION

While there exist many reactions that can alter contamination concentrations in groundwater, adsorption onto the soil matrix is one of the dominant mechanisms. The concept of the isotherm is used to relate the amount of contaminant adsorbed by the solids S to the concentration in solution, C. One of the most commonly used forms is the Freundlich isotherm,

$$S = K_d C^b \tag{5.1}$$

where S is the mass of solute adsorbed per unit bulk dry mass of porous media, K_d is the distribution coefficient, and b is an experimentally derived coefficient. If $b = 1$, Equation (5.1) is known as the linear isotherm and is incorporated into the 1-D advective-dispersion equation through $R = [1 + (\varrho_b/n)K_d] = $ retardation factor, which has the effect of retarding the adsorbed species relative to the advective velocity of the groundwater (Figure 5.4 and 5.5). The retardation factor can be a useful tool for the case of linear isotherms with fast, reversible adsorption. Retarded fronts in one dimension can be derived from conservative fronts by dividing velocity and longitudinal dispersivity by R. Typical values of R for organics often encountered in field sites range from 1 to 50. Pickens and Lennox, 1976 [25] present one of the first graphic displays of the effects of K_d and dispersivity values on observed front locations in two dimensions. Roberts et al., 1986 [26] found retardation factors ranging from 1.5 to 9.0 for organics in the natural gradient test at the Borden landfill in Canada.

The use of the distribution coefficient assumes that partitioning reactions between the solute and soil are very fast relative to the rate of groundwater flow. Thus, it is possible for nonequilibrium fronts to occur that appear to migrate faster than retarded fronts, which are at equilibrium. These complexities involve other rate kinetic factors beyond the scope of simple models discussed in this chapter. Because of these effects, the plume of a reactive contaminant expands more slowly and the concentration is less than that of an equivalent nonreactive contaminant. Unfortunately, this retarding effect can significantly increase the cleanup time of a pump-and-treat system.

BIODEGRADATION

In addition to the transport processes discussed above, other information may need to be collected relating to biodegradation processes that may be occurring at a site. A minimum list of data required to evaluate the potential for biodegradation of organics includes (1) characterization of organisms in the subsurface, (2) analysis for dissolved oxygen and nutrients in sufficient concentrations required for the biological process to occur, and (3) analysis for potential transformation products (degradation compounds). The above steps are important in order to estimate natural degradation and to determine if bioremediation, which includes the injection of oxygen or other nutrients into the subsurface to stimulate existing microbes, is a possible remedial alternative. In situ biodegradation as a remediation method is described in more detail in Chapter 6.

SITE CHARACTERIZATION AND MONITORING

Subsurface conditions can be studied only by indirect techniques or by using point well or borehole data. Common data collection methods in the subsurface include geophysical techniques, drilling, groundwater sampling, soil sampling, and aquifer tests. References on monitoring wells include Scalf et al., 1981 [27]; Driscoll, 1986 [28]; and Campbell and Lehr, 1973 [29]; references on geophysical techniques include Dobrin, 1976 [30]; Keys and MacCary, 1971 [31]; Stewart et al., 1983 [32]; and Kwader, 1986 [33]. Choice of appropriate sampling methods depends on the overall objectives of the project. Throughout the field study, it is essential to document all well construction details, sampling episodes, and field tests in order to arrive at an accurate evaluation of the entire site. A full understanding of the hydrogeology and extent of contamination at a site are important to a successful remedial design. Formulating detailed vertical stratigraphy ensures that wells are sited to a proper depth and stratigraphic layer so that any potential cross contamination is minimized.

Methods for determining hydraulic properties of subsurface units primarily consist of aquifer tests such as pump tests or slug tests. In a pump test, a well is pumped at a constant rate and water-level responses are measured in surrounding wells. Solutions are available for estimating aquifer parameters based on the pump rate and the response, which could be either drawdown or recovery (Ferris et al., 1962 [34]; Kruseman and DeRidder, 1976 [35]; Fetter, 1988 [2]). Essential equations and examples of pump tests are described in Chapter 2 and will not be repeated here.

The slug test method involves inducing a rapid water-level change within a well and measuring the head changes as the water level in the well returns to its initial level. The initial water-level change can be induced by either introducing or withdrawing a volume of water or displacement device into or out of the well. The rate of recovery is related to the hydraulic conductivity of the surrounding aquifer material (Hvorslev, 1951 [36]; Cooper et al., 1967 [37]; Papadopulos et al., 1973 [38]; Bouwer and Rice, 1976 [39]). The advantage of a slug test over a pump test is that little or no contaminated water will be produced. Unfortunately, slug tests measure the response in only a small volume of the permeable media, whereas aquifer tests measure the response in a much larger volume. Standard texts and Chapter 2 provide details on analyzing slug test data (Freeze and Cherry, 1979 [1]; Fetter, 1988 [2]).

To determine flow directions and vertical and horizontal gradients, water levels must be measured and converted to elevations relative to a datum, usually mean sea level. Water-level measurements may be taken by several different means including chalk and tape, electrical water-level probe, and pressure transducer as discussed in Fetter, 1988 [2] and Streltsova, 1988 [40]. Horizontal gradients are determined using water-level data from wells that are screened into the same aquifer layer and/or at the same elevation but separated areally. The water elevations are often plotted on a map as contours so that flow direction can be established as shown in Figure 5.1.

Vertical gradients are determined using water-level data from wells in the same location but screened to different elevations. The gradient is the difference in water levels divided by the distance between the measurement locations. Vertical gradients often occur across confining beds of clays or silts. Because water levels often yield a complex three-dimensional surface, care must be taken in computing the hydraulic gradient, which determines the direction of flow. Figure 5.6 shows vertical differences in heads between an unconfined and a leaky confined aquifer unit. Such vertical paths of flow must be carefully evaluated prior to any remedial plan.

For fractured media and karst formations, site characterization and remediation designs are even more difficult. The analysis can be carried out assuming that the fractured media can be represented by an elementary volume with an average set of aquifer parameters (Snow, 1968 [41]). More

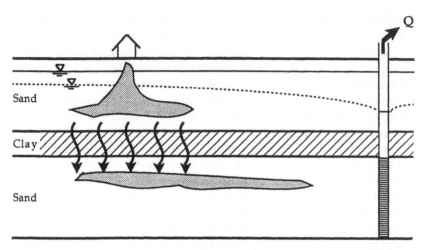

Figure 5.6 Vertical flow in groundwater.

recently, efforts have emphasized an approach based on flow and transport in individual fractures, but the distribution, continuity, and interconnectivity of fractures govern the response [40]. Techniques such as fracture trace analysis (Lattman and Parizek, 1964 [42]) and the use of geophysical instrumentation may be useful for locating the more permeable zones, where contaminants are most likely to be located and where extraction wells should be placed. For more detailed discussion on flow in the special heterogeneous conditions of fractured media, see Streltsova, 1988 [40]; for karst formations, see Bogli, 1980 [43].

ANALYSIS OF PLUME MOVEMENT

Once a site has been accurately characterized, the first step in data interpretation is making preliminary calculations such as using the hydraulic gradient, hydraulic conductivity, and porosity in Darcy's equation to estimate advective transport. Next, one may compare these velocity calculations with estimates of mean plume movement. If the two are not at all comparable, this could indicate uncertainty in the source release rate or location, or that processes such as sorption or biodegradation are important. There are numerous analytical and numerical tools that can be used to interpret groundwater plume data, including geostatistical analyses and computer modeling.

GEOSTATISTICS

Methods such as kriging can be used to quantify the spatial variability inherent in the hydraulic conductivity field or concentration plume of an aquifer (Journel and Huijbregts, 1978 [44]; Englund and Sparks, 1988 [45]). De Marsily, 1986 [14] presents a concise review of geostatistical approaches in hydrogeology. Having measured a variable at a set of points in space, how do we estimate the value of the variable at all other locations in order to produce a contour map? Kriging is an optimal estimation method that can be used to estimate transmissivity, heads, and concentrations based on spatial measurements. Co-kriging is an estimation technique useful when two or more variables that are correlated are measured in the field and can be estimated together (DeMarsily, 1986 [14]). Cooper and Istok, 1988 [47] present an application of geostatistical analysis to groundwater quality.

GROUNDWATER MODELS

Groundwater flow and transport modeling performed during the remedial investigation can be a powerful tool to estimate plume movement and response to various remedial schemes. Flow and contaminant transport models should be calibrated to a measured plume of contamination to the extent possible. However, caution should be used when applying models at Superfund sites because there is uncertainty whenever subsurface movement is modeled, particularly when the results of the model are based on estimated parameters. The purposes of modeling groundwater flow include the following:

(1) To predict concentrations of contaminants at receptor points
(2) To estimate the effect of source-control actions on remediation
(3) To guide the placement of monitoring wells and hydrogeologic characterization when the remedial study is conducted in phases
(4) To evaluate expected remedy performance during the feasibility study so that the effects of restoration can be predicted

Various analytical and numerical models are available to predict solute contaminant concentrations in one, two, and three dimensions. They can also be used to predict remedial alternative performance (Table 5.2). The models vary in the number of simplifying assumptions that must be made, the cost of running the model, and the level of effort needed. Most numerical models incorporate more information and require more data and expertise to run than simpler analytical models. However, analytical models are not as rigorous in treating variable aquifer parameters. Regardless of the complexity of the model, however, representative input data must be used to obtain reliable results, and the results of the models must be interpreted with care.

The determination of whether or not to use modeling and the level of effort that should be expended is made on the basis of the objectives of the

TABLE 5.2. Summary of Codes.

Code Name	Developer	Method	Comments
Flow			
PLASM (2-D)	T. A. Prickett & Assoc.	finite difference	PLASM stands for the Prickett-Lonnquist Aquifer Simulation Model. The finite difference equations are solved using a form of the iterative alternating direction implicit procedure (IADIP).
MODFLOW (3-D) (MODPATH)	USGS	finite difference	MODFLOW is a MODular three-dimensional finite difference FLOW code developed by the U.S. Geological Survey. The code permits the user to select a series of packages (or modules) to be used during a given simulation.
Transport			
BIO1D (1-D)	GeoTrans, Inc.	finite difference	BIO1D is designed to evaluate one-dimensional transport of reactive dissolved species that undergo adsorption and degradation. The model is ideally suited for deciding which transport processes are important in a given problem.
RNDWALK (2-D)	T. A. Prickett & Assoc.	finite difference with random walk	RNDWALK is the RaNDom-WALK model, originally developed by the Illinois State Water Survey, that simulated 2-D contaminant transport.

TABLE 5.2. (continued).

Code Name	Developer	Method	Comments
USGS MOC (2-D)	USGS	finite difference/MOC	The flow and transport code, MOC or Method Of Characteristics, solves contaminant transport in two dimensions and was developed by the USGS.
BIOPLUME II (2-D)	Rifai et al.	finite difference/MOC	The MOC code has also been extended to treat variable density fluids and to allow parallel computation of an organic plume subject to biodegradation and an oxygen plume.
SWIFT II (3-D)	GeoTrans, Inc.	finite difference	The Sandia Waste-Isolation Flow and Transport model, SWIFT II, is a three-dimensional finite-difference model for fluid flow, heat, variable density-brine, and trace-level radionuclide transport.
HST3D (3-D)	USGS	finite difference	HST3D is a Heat- and Solute-Transport code for 3-D simulations, which couples mesh-centered finite difference equations for flow, heat, and solute transport including linear equilibrium adsorption and first-order decay. Confined and unconfined aquifers under transient conditions can be represented.

(continued)

99

TABLE 5.2. (continued).

Code Name	Developer	Method	Comments
PATH3D (3-D)	Papadopulos & Assoc.	Runge-Kutta	PATH3D uses a fourth-order Runge-Kutta numerical tracking solution that allows a particle to take several small tracking steps while moving across a finite difference block.
PRZM (1-D)	U.S.EPA	compartment	PRZM or the Pesticide Root Zone Model is a time-dependent, compartment model that simulates chemical movement under unsaturated conditions in the plant root and vadose zones.
SESOIL (1-D)	U.S.EPA	compartment	SESOIL is a SEasonal compartment model for a SOIL column the EPA designed to provide multiple-year simulations of water, chemical, and sediment transport using a monthly time step.
SUTRA (3-D)	USGS	finite difference finite element	SUTRA (Saturated-Unsaturated TRAnsport) simulates two-dimensional, density-dependent flow and transport of either a dissolved solute or thermal energy under variably saturated conditions. SUTRA was developed by the U.S. Geological Survey to provide investigators with a tool to evaluate the importance of concentration or temperature on density-dependent flow.

modeling, the ease with which the subsurface can be conceptualized mathematically, and the availability of data. Field data are collected to characterize the variables that govern the hydrologic and contaminant response of the site in question. Estimates based on literature values or professional judgment are frequently used as well.

Models such as the three-dimensional, finite-difference flow code MODFLOW (McDonald and Harbaugh, 1984 [48]) and the semianalytical flow code RESSQ (Javandel et al., 1984 [49]) can be used to simulate flow patterns and changes resulting from the operation of a pump-and-treat system. Other models are available to analyze contaminant transport such as the popular two-dimensional MOC model from Konikow and Bredehoeft, 1978 [50] and Bioplume II from Rifai et al., 1988 [51]. A detailed modeling exercise simulating the pump-and-treat remediation of an aquifer is described later in this chapter.

SELECTION OF REMEDIAL ALTERNATIVES

Once a site has been well characterized for hydrogeology and contaminant concentrations, alternatives for control and remediation can be selected and combined to provide an overall strategy for cleanup. Choosing a remedial technology is a function of the contaminant, site characteristics, and the location of the contaminant with respect to the water table. Hydraulic conductivity or transmissivity of a formation is the most important parameter since pump rates and velocities are directly related. The reactivity or biodegradability of the contaminant is vital for determining whether an in situ treatment process will work. If pure product exists at or near the water table in the form of separate phase fluid, the problem of removal may be greatly complicated as described in Chapter 8. Thus, depending on the situation, it may be necessary to combine a pumping system with other techniques (bioderemediation, soil venting, skimming of product) in order to complete remediation in the saturated and vadose zones.

Several alternatives for handling groundwater contamination problems include: (1) containment, (2) excavation, (3) pumped removal of product or contaminated water, (4) in situ treatment (chemical or biological), and (5) vacuum extraction.

The containment option is designed to control the spread of contaminants in the subsurface by the use of physical containment methods or hydrodynamic barriers. Hydrodynamic controls usually involve some pumping of groundwater via a series of wells surrounding or in the plume.

Physical containment measures are designed to isolate contaminated soil and groundwater from the local environment and to minimize any threat to human health. Isolation techniques for the surface and subsurface include excavation and removal of the contaminated soil and groundwater, barriers to groundwater flow, and surface water controls. Historical barrier approaches include slurry walls, grout curtains, sheet piling, and liners or geomembranes. Ehrenfeld and Bass, 1984 [52]; Canter and Knox, 1986 [53]; Thomas et al., 1987 [54]; and Mercer et al., 1990 [55] provide very thorough reviews of groundwater pollution controls and remedial methods. An evaluation of actual extraction remedies used at sites in the U.S. has been completed by EPA, 1989 [56].

Excavation methods usually involve digging of a pit to remove the soil, or pumping wells are installed to control the plume, and the excavated soil is transported to a secured site, such as a landfill or surface impoundment, for disposal (Ehrenfeld and Bass, 1984 [52]). The groundwater is pumped out and can be treated using a variety of techniques. An obvious difficulty associated with excavation and removal is that total removal of subsurface soil and groundwater may be impossible when the contamination extends deep beneath the surface or the contaminants lay below an immobile facility.

Physical barriers used to prevent the flow of groundwater include slurry walls, grout curtains, sheet piling, various man-made liners, and block displacement (Ehrenfeld and Bass, 1984 [52]; Canter and Knox, 1986 [53]). The use of these methods at hazardous waste sites has resulted in serious problems of leakage through and around the barriers; therefore, hydraulic controls are generally preferred for most cases.

HYDRAULIC CONTROLS AND CONTAMINANT WITHDRAWAL SYSTEMS

Hydraulic control of groundwater contamination can be designed to generally lower the water table to prevent discharge to a river or lake, to reduce the rate of migration by dewatering the waste, or to confine the plume to a potentiometric low created by a combination of pumping and injection wells. Maintenance on wells and pumps is particularly important for this method, and as groundwater levels change, the system design may have to respond accordingly. On-site treatment works may be necessary to handle any contaminated water pumped by the wells.

Four main alternatives for removing contaminants from the subsurface may all be part of an overall strategy for site remediation. The most popular and successful methods include excavation, interceptor systems, soil venting, and pumping/injection. It is not uncommon to find some or all of these methods being used at a single site, depending on the mix of chemicals and the hydrogeology of the site.

Interceptor systems use drains, a line of buried perforated pipe, and/or trenches, and an open excavation usually backfilled with gravel to collect contaminated groundwater close to the water table. These systems operate as an infinite line of wells near the shallow water table and are efficient at removing contamination near the surface. Trenches are often used to collect nonaqueous phase liquids (NAPLs) like crude oil or gasoline, which are light and tend to move near the capillary fringe just above the water table.

Soil venting removes volatile organic contaminants from the unsaturated zone using a vacuum pumping approach. When operated properly, soil venting or vapor extraction can be one of the most cost-effective remediation methods for soils contaminated with volatile compounds such as gasoline or solvents. The design of a soil venting system is complex, and design criteria include the number of wells, well spacing, well location, construction, and vapor treatment systems (Johnson et al., 1990 [57,58]). Soil venting is presented in more detail in Chapter 7.

Pumping wells are used to extract water from the saturated zone by creating a capture zone for migrating contaminants. A major problem is the proper treatment and disposal of the contaminated water. On-site treatment facilities are usually required before water can be reinjected to the aquifer or released to the surface system. The number of wells, their locations, and the required pumping rates are the key design parameters of interest, and methods of analysis are described in more detail in Chapter 2 in the book.

Pumping water containing dissolved contaminants can be addressed using standard well mechanics and capture zone theory which is well-understood. If the hydraulic conductivity is too low or the geology is very complex and heterogeneous, then pumping may not be a feasible alternative for hazardous waste cleanup. If the hydrogeology is conducive to an injection pumping system, then several design approaches can be used to develop an efficient and reliable system for contaminant removal.

Pilot scale systems or small field demonstration projects have been used at a number of sites to evaluate the pump rates and the placement of wells in a small area of the site, before expanding to the entire site. In this way, operational policies, mechanical problems, and costs can be evaluated before attempting the larger cleanup. Careful monitoring of the system is the key to understanding how the injection pumping pattern will respond over time.

Javendel and Tsang, 1986 [59] developed a useful analytical method for the design of recovery well systems, based on the concept of a capture zone (Figure 5.7). The capture zone for a well depends on the pumping rate and the aquifer conditions. Ideally, the capture zone should be somewhat larger than the plume to be cleaned up, and thus wells can be added until

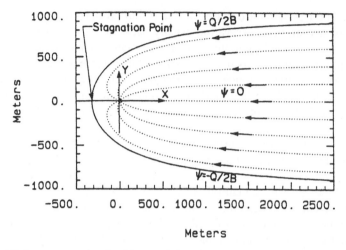

Figure 5.7 Capture zones (reprinted by permission of the *Journal of Ground Water*, copyright 1986, worldwide rights reserved).

sufficient pumping capacity is provided to create a useful capture zone. However, with more wells, some contaminants may pass between the wells, and well spacing becomes an important parameter as well as pumping rate. The greater the pumping rate, the larger the capture zone, and the closer the wells are placed, the better the chance of complete plume capture. Overall, the method from Javendel and Tsang minimizes the pumping injection rates through a proper choice of well location and distance between wells.

This method is extremely useful but cannot be applied to every design problem encountered in practice. There are assumptions built into the formulation such as constant aquifer transmissivity, fully penetrating wells, no recharge, and isotropic hydraulic conductivity, which have to be satisfied by the field problem or the method will not necessarily yield a correct result. Actual field sites where boundary conditions and site variabilities are important issues may require analysis using numerical models as described in the next section.

MODELING OF TCE RECOVERY IN A SHALLOW AQUIFER

INTRODUCTION

Combined field and computer studies at a given site can be used to develop remedial schemes for addressing the problem of groundwater contamination. This study demonstrates some of the capabilities of ground-water contaminant modeling applied to plume delineation and aquifer restoration (Freeberg et al., 1987 [60]). The USGS MOC Solute Transport Model (Konikow and Bredehoeft, 1978 [61]) is applied to an industrial site where a mixture of solvents (primarily trichloroethylene, TCE) leaked from an underground storage tank to the underlying sand aquifer. The model in its two-dimensional form for steady-state flow is capable of simulating the horizontal movement of a contaminant plume under both ambient flow conditions and conditions produced by remedial activity. For this reason, the model is well-suited to application at industrial sites where contaminant movement and recovery systems are monitored.

HYDROGEOLOGIC SETTING

Investigation of near-surface hydrogeologic conditions showed that the site is situated on sand deposits extending 10 to 35 ft (3 to 10.7 m) below ground surface. Geologic logs of monitoring wells at the site showed that continuous clay layers ranging in thickness from 20 to 100 ft (6 to 30 m) separate the surface sands from the Lower Chicot aquifer, which is pumped for municipal water supply. Because contaminant migration occurred primarily in a lateral direction within a shallow, confined, relatively homogeneous sand aquifer, the site provided an ideal opportunity for application of the USGS Solute Transport Model.

Fifteen monitoring wells were placed in the vicinity of the storage tank and across the predicted path of the migrating plume. Boreholes and PVC well casings were screened over the entire saturated sand thickness, which was both underlain and overlain by confining clay layers.

Monitoring of the system, using the configuration of wells shown in Figure 5.8, began after the leaking tank had been removed and continued through the installation and operation of a four-well recovery system. Water-level measurements and analyses of samples taken from the wells over a 1.5-year period provided an extensive data base with which to calibrate the computer model. Because the wells were fully screened through the entire thickness of the saturated sand, TCE concentrations in groundwater samples taken from the wells represented vertically averaged concentrations.

Water-level measurements, as well as data from a constant-discharge pumping test, provided information on the hydraulic regime. Groundwater movement in the system occurs as an essentially linear flow field in a southeastern direction. Some bending of flow lines occurs in the vicinity of a ditch that bounded the site along its southern edge.

The groundwater velocity in the sand aquifer was relatively high and consequently the movement of contaminants was readily detectable during

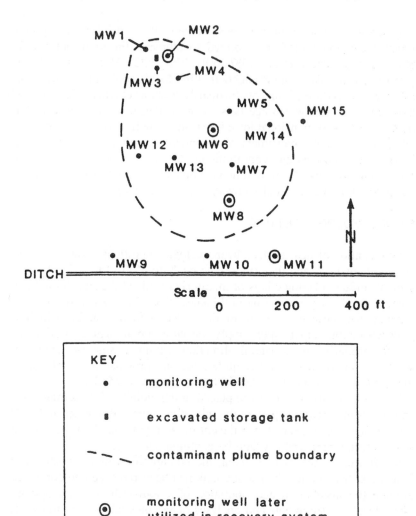

Figure 5.8 Monitoring wells at TCE site (2) (reprinted by permission of the *Journal of Ground Water*, worldwide rights reserved, copyright 1987).

the monitoring period. Results from the pumping test indicated that the transmissivity of the sand unit was approximately 2220 ft²/day (206 m²/day). With an average saturated thickness of 25 ft (7.6 m), the hydraulic conductivity of the aquifer was calculated to be 1.03×10^{-3} ft/sec (3.14×10^{-2} cm/sec). A storage coefficient of 0.0075 was determined from pumping-test data by Theis analysis, indicating that the aquifer was confined. Assuming a typical value for effective porosity (0.3) and an average hydraulic gradient of 0.002, seepage velocity at the site was found to be approximately 200 ft/yr (60 m/yr).

Groundwater samples collected during the monitoring period were analyzed on a gas chromatograph equipped with a flame ionization detector and a purge and trap apparatus to achieve sensitivity of low concentrations at the μg/l level. The migration of trichloroethylene (TCE), the compound observed in the highest concentrations, was addressed in the application of the USGS Solute Transport Model. This pollutant was modeled as a conservative tracer because it undergoes negligible adsorption (Schwartzenbach and Giger, 1985 [62]) or biodegradation in the system of interest (Bouwer et al., 1981 [63]).

HISTORY OF CONTAMINATION

Chemical data collected during the monitoring period represented the only record of contaminant movement on the site. The underground storage tank was in use fifteen years before it was excavated and groundwater monitoring began, but the history of leakage during this period is unknown. The 5000-gallon (19-m³) capacity concrete tank contained primarily a mixture of varying proportions of TCE and toluene. The tank was repeatedly filled and drained during the service period. Consequently, the release of contaminants, through periodic accidental spillage and from continuous leakage through the vinyl seams of the tank, which had gradually dissolved away, occurred at an unknown rate. Chemical data from the midway point in the monitoring period allowed delineation of the areal extent of contamination (Figure 5.8). Calculation of the toal volume of contaminated water (12.3 × 10⁶ gallons or 46.0 × 10³ m³) and determination of the average concentration of contaminant exhibited in the plume at that time (176 μg/l TCE) allowed the total mass of contaminant to be estimated. It was estimated that the leaking storage tank had released a total load of 8 kg of TCE.

REMEDIAL ACTIVITY

Remedial activity at the industrial site began approximately one year after the solvent tank was excavated and monitoring had begun. The recovery system was designed (1) to produce an area of drawdown sufficient to encompass the contaminant plume; (2) to minimize the total flow rate and, thereby, reduce the volume of contaminated groundwater to be treated; and (3) to minimize installation and operating costs.

The most cost-effective recovery system was found to be one that employed monitoring wells already in existence. The recovery system consisted of four monitoring wells which were to be pumped continuously at a design flow rate of 30 gpm (160 m³/day) each.

APPLICATION OF THE MODEL

The USGS Solute Transport Model, a widely used, well-documented computer code, was applied to the site in its two-dimensional form for steady-state flow conditions. The model represents the flow field using a rectangular, block-centered finite-difference grid. It simulates the movement of a contaminant plume by computing changes in concentration caused by advective transport, hydrodynamic dispersion, and mixing from fluid sources.

The particle-tracking method used in the USGS Solute Transport Model is capable of delineating sharp contaminant gradients. Therefore, it is especially appropriate in this application, where transport of TCE in a homogeneous sand is advection-dominated. In this case study, dispersion was not a significant transport process in the pumped system, and saturated thickness did not vary significantly over the site.

CONTAMINANT PLUME PREDICTION

The study area was simulated using a 20 × 35 cell grid with each cell representing a 50 × 50 ft (15 × 15 m) area. Five of the fifteen wells monitored at the site were designated as "observation wells" in the model. Actual water-quality data from wells MW-4, 5, 12, 14, and 7 provided the most information about movement of the plume and distribution of contaminants. Predicted data from observation wells were useful in calibrating the model predictions to observed data (Figure 5.9).

Other parameters describing the hydrogeologic regime at the site were adjusted during calibration of the model. Ranges of values were used in calibration for parameters such as effective porosity, longitudinal dispersivity, ratio of transverse to longitudinal dispersivity, ratio of transmissivity in the y-direction to transmissivity in the x-direction (ratio of T_{yy} to T_{xx}), saturated thickness, and background contaminant concentration. The length of the injection period, which simulated the period of leakage prior to removal of the tank, and the concentration of TCE in the injection fluid were chosen so that a total mass of 8 kg of TCE was released into the aquifer.

MODELING RESULTS

Simulation of the recovery system followed simulation of the monitoring period described above. Four recovery wells corresponding to MW-2, 6, 8,

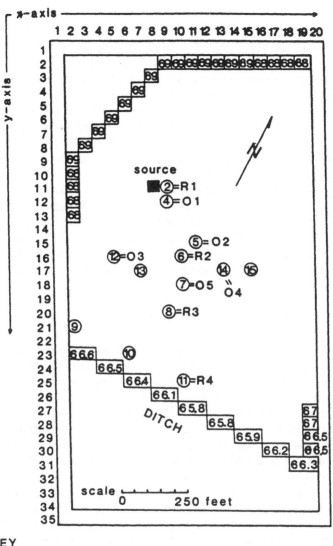

KEY

68	constant head node with potentiometric value indicated in box	R 1–4 = recovery well
②	monitoring well	O 1–5 = designated observation well in simulations

Figure 5.9 Model setup (reprinted by permission of the *Journal of Ground Water*, worldwide rights reserved, copyright 1987).

and 11 were specified in the model (Figure 5.9). Pumping rates corresponding to the average pumping rates actually applied at the site also were specified. The recovery wells were turned on at the start of the simulation and allowed to operate continuously until the end of the simulation period.

The first part of the modeling study addressed simulation of the plume of TCE created by the leaking storage tank. Calibration of the hydraulic gradient produced an excellent fit between predicted and observed head distributions. Calibration of the model, which required more than 100 computer runs, continued until the error between the predicted and measured contaminant plumes was minimized.

The contaminant loading chosen in the final calibration of the model, a half-year injection of TCE at a concentration of 1500 μg/l and a rate of 0.012 cfs (3.4 $\times$ 10^{-4} m^3/s), produced the contaminant plume shown in Figure 5.10 at the end of the one-year monitoring period. To illustrate the fit of the prediction, concentrations of TCE actually observed on the site at the same point in time also are indicated at monitoring well locations in Figure 5.10.

RESULTS OF RECOVERY SIMULATIONS

Simulation of the recovery system showed that operation of the four wells produced a marked decrease in the extent of the plume and in the predicted concentrations of TCE. Figure 5.11 show contaminant distributions predicted by the model after 0.5, 1.0, 1.5, and 2.0 years of pumping. At 2.0 years in the simulation, a mass balance error of 0.46% was computed, again indicating good numerical accuracy in the solution.

Comparisons of actual chemical and water-level data to values predicted in the simulation also provide a good check on the accuracy of the model.

Simulated predictions about the effectiveness of the recovery system could be tested at two points in time. In June, 1984, 0.16 years after the start-up of the recovery system, analyses for TCE were conducted on samples from five of the monitoring wells. A comparison of this data to contaminant plume concentrations predicted after 0.16 years is provided.

In September, 1984, 0.47 years after pumping began, nine of the monitoring wells were sampled, provided a second check on the accuracy of the recovery system simulation. When the concentrations of TCE measured at these nine monitoring wells are totaled, the chemical data show a 90% reduction in the total concentration of TCE measured at the same nine wells prior to pumping. The simulation predicts an 89% reduction in TCE concentrations at the nine wells during the 0.47 years of pumping, suggesting that the simulation of the recovery system is fairly accurate (Figure 5.11).

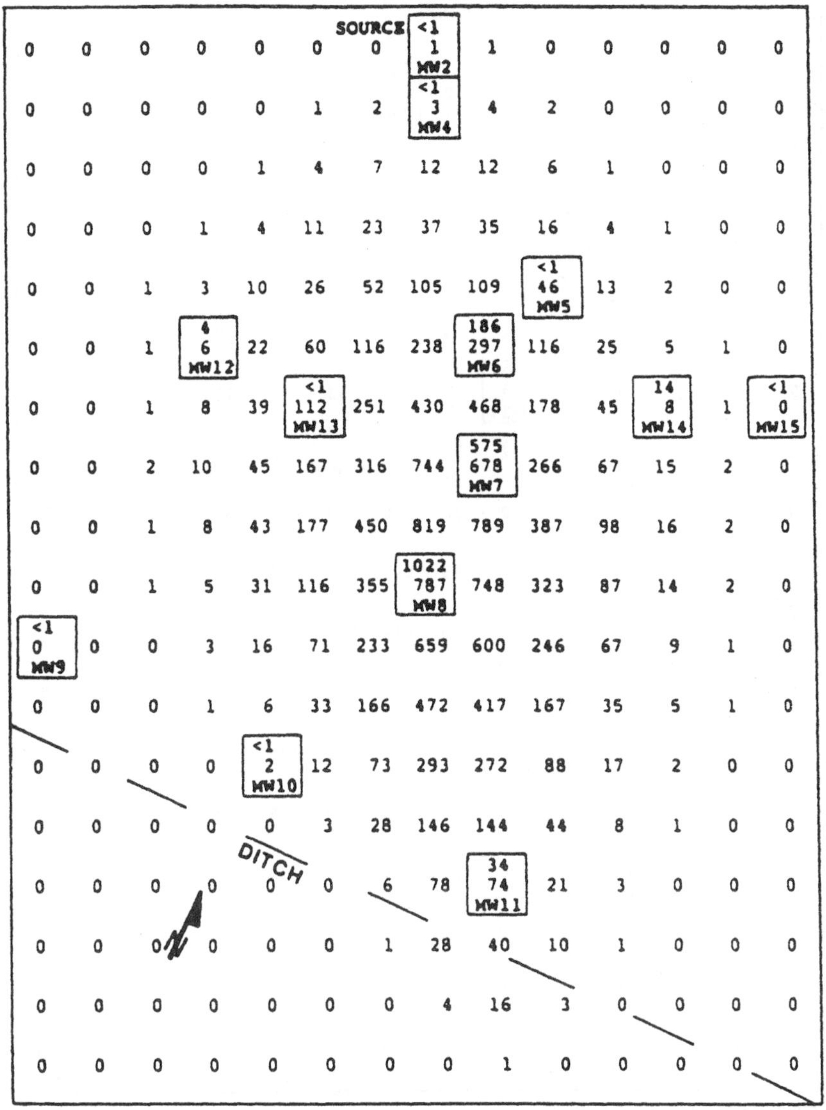

						SOURCE	<1 1 MW2	1					
0	0	0	0	0	0	0	<1 / 1 / MW2	1	0	0	0	0	0
0	0	0	0	0	1	2	<1 / 3 / MW4	4	2	0	0	0	0
0	0	0	0	1	4	7	12	12	6	1	0	0	0
0	0	0	1	4	11	23	37	35	16	4	1	0	0
0	0	1	3	10	26	52	105	109	<1 / 46 / MW5	13	2	0	0
0	0	1	4 / 6 / MW12	22	60	116	238	186 / 297 / MW6	116	25	5	1	0
0	0	1	8	39	<1 / 112 / MW13	251	430	468	178	45	14 / 8 / MW14	1	<1 / 0 / MW15
0	0	2	10	45	167	316	744	575 / 678 / MW7	266	67	15	2	0
0	0	1	8	43	177	450	819	789	387	98	16	2	0
0	0	1	5	31	116	355	1022 / 787 / MW8	748	323	87	14	2	0
<1 / 0 / MW9	0	0	3	16	71	233	659	600	246	67	9	1	0
0	0	0	1	6	33	166	472	417	167	35	5	1	0
0	0	0	0	<1 / 2 / MW10	12	73	293	272	88	17	2	0	0
0	0	0	0	0	3	28	146	144	44	8	1	0	0
0	0	0	0	0 DITCH	0	6	78	34 / 74 / MW11	21	3	0	0	0
0	0	0	0	0	0	1	28	40	10	1	0	0	0
0	0	0	0	0	0	0	4	16	3	0	0	0	0
0	0	0	0	0	0	0	0	1	0	0	0	0	0

Boxes indicate monitoring well locations.
Bold figures represent observed concentrations of TCE (μg/l), provided for comparison to values predicted across the grid.

0   50   100 feet

Figure 5.10 Predicted plume (reprinted by permission of the *Journal of Ground Water*, world-wide rights reserved, copyright 1987).

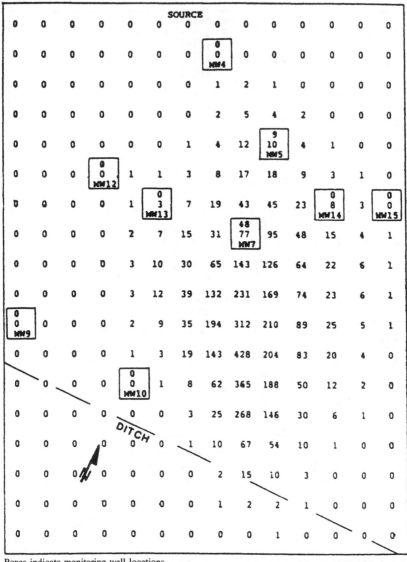

Boxes indicate monitoring well locations.
Bold figures represent observed concentrations
of TCE (μg/l), provided for comparison to
values predicted across the grid.

Figure 5.11 Recovery system results at 0.5 years (reprinted by permission of the *Journal of Ground Water*, worldwide rights reserved, copyright 1987).

Despite variations between the predicted results and actual observations, simulation of the contaminant plume and recovery system at the industrial site provides a useful prediction of plume migration and of the efficiency of contaminant removal during remedial activity. The model shows that a recovery period on the order of 2.0 years is necessary to reduce concentrations of TCE to 6 μg/l or less. The modeling exercise also indicates gaps in the data base concerning the existence of spatial variations in transmissivity or other heterogeneities in the system. Monitoring data for calibration of the model were available from only four wells located within the measured extent of the plume. Predictions made by the model at the four wells matched actual observations, at three points in time, before and after start-up of the recovery system.

REFERENCES

1 Freeze, R. A. and J. A. Cherry. 1979. *Groundwater*, Prentice-Hall, Englewood Cliffs, NJ.

2 Fetter, C. W. 1988. *Applied Hydrogeology*, Merrill Publishing Company, Columbus, OH, p. 592.

3 U.S. Environmental Protection Agency. 1986. "RCRA Ground-Water Monitoring Technical Enforcement Guidance Document," OSWER-9950.1, Washington, D.C.

4 U.S. Environmental Protection Agency. 1987. "Handbook Ground Water," EPA/625/6-87/016, Cincinnati, OH, p. 212.

5 Cartwright, K. and J. M. Shafer. 1987. "Selected Technical Considerations for Data Collection and Interpretation – Groundwater," in *National Water Quality Monitoring and Assessment*, Washington, D.C.

6 Barcelona, M. J., J. P. Gibb and R. A. Miller. 1983. "A Guide to the Selection of Materials for Monitoring Well Construction and Ground-Water Sampling," Illinois State Water Survey Contract Report No. 327, USEPA-RSKERL, EPA/600/52-84/024, U.S. Environmental Protection Agency.

7 Barcelona, M. J., J. P. Gibb, J. A. Helfrich and E. E. Garske. 1985. "Practical Guide for Ground-Water Sampling," Illinois State Water Survey Contract Report No. 374, USEPA-RSKERL under cooperative agreement CR-809966-01, U.S. Environmental Protection Agency, Ada, OK.

8 Mackay, D. M. and J. A. Cherry. 1989. "Groundwater Contamination: Pump-and-Treat Remediation," *Environ. Sci. Technol.*, 23:6:630–636.

9 Anderson, M. P. 1979. "Using Models to Simulate the Movement of Contaminants through Ground Water Flow Systems," *CRC Critical Rev. Environ. Control*, Chemical Rubber Co., 9, 96.

10 Anderson, M. P. 1984. "Movement of Contaminants in Groundwater: Groundwater Transport – Advection and Dispersion," *Studies in Geophysics, Groundwater Contamination*, National Academy Press, Washington, D.C., pp. 37–45.

11 Wang, H. F. and M. P. Anderson. 1982. *Introduction to Groundwater Modeling, Finite Difference and Finite Element Methods*, W. H. Freeman and Company, San Francisco, CA, p. 237.

12 Bear, J. 1979. *Hydraulics of Ground Water*, McGraw-Hill, New York, NY.

13 Bedient, P. B., R. C. Borden and D. I. Leib. 1985. "Chapter 28: Basic Concepts for Ground Water Modeling," *Ground Water Quality*, C. H. Ward, W. Giger, and P. L. McCarty, eds., John Wiley and Sons, pp. 512–531.

14 DeMarsily, G. 1986. *Quantitative Hydrogeology*, Academic Press, Inc., Harcourt Brace Jovanovich, Publishers, Orlando, FL, p. 440.

15 Nelson, R. W. 1978. "Evaluating the Environmental Consequences of Ground-water Contamination, Parts 1–4," *Water Resour. Res.*, 14:3:409–450.

16 Charbeneau, R. J. 1981. "Groundwater Contaminant Transport with Adsorption and Ion Exchange Chemistry: Method of Characteristics for the Case without Dispersion," *Water Resour. Res.*, 17:3:705–713.

17 Charbeneau, R. J. 1982. "Calculation of Pollutant Removal during Groundwater Restoration with Adsorption and Ion Exchange," *Water Resour. Res.*, 18:4:1117–1125.

18 Tsang, C. F. and I. Javendahl. 1986. "Capture-Zone Type Curves: A Tool for Aquifer Cleanup," *Ground Water*, 24:5:616–625.

19 Gelhar, L. W., A. L. Gutjahr and R. L. Naff. 1979. "Stochastic Analysis of Macrodispersion in a Stratified Aquifer," *Water Resour. Res.*, 15:6:1387–1397.

20 Gelhar, L. W., A. Montoglou, C. Welty and K. R. Rehfeldt. 1985. "A Review of Fieldscale Physical Solute Transport Processes in Saturated and Unsaturated Porous Media," Final Proj. Rep EPRI EA-4190, Elec. Power Res. Inst., Palo Alto, CA.

21 Freyberg, D. L. 1986. "A Natural Gradient Experiment on Solute Transport in a Sand Aquifer, (2) Apatial Moments and the Advection and Dispersion of Nonreactive Tracers," *Water Resour. Res.*, 22:13:2031–2046.

22 Dagan, G. 1984. "Solute Transport in Heterogeneous Porous Formations," *J. Fluid Mech.*, 145:151–177.

23 Mackay, D. M., D. L. Freyberg, P. V. Roberts and J. A. Cherry. 1986. "A Natural Gradient Experiment on Solute Transport in a Sand Aquifer, (1) Approach and Overview of Plume Movement," *Water Resour. Res.*, 22:13:2017–2029.

24 Sudicky, E. A. 1986. "A Natural Gradient Experiment on Solute Transport in a Sand Aquifer: Spatial Variability of Hydraulic Conductivity and Its Role in the Dispersion Process," *Water Resour. Res.*, 22:13:2069–2082.

25 Pickens, J. F. and W. C. Lennox. 1976. "Numerical Simulation of Waste Movement in Steady Groundwater Flow Systems," *Water Resour. Res.*, 12:2:171–180.

26 Roberts, P. V., M. N. Goltz and D. M. Mackay. 1986. "A Natural Gradient Experiment on Solute Transport in a Sand Aquifer, (4), Sorption of Organic Solutes and Its Influence on Mobility," *Water Resour. Res.*, 22:13:2047–2058.

27 Scalf, M. R., S. F. McNabb, W. I. Dunlap, R. L. Cosby and J. Fryberger. 1981. "Manual of Groundwater Quality Sampling Procedures," Robert S. Kerr Environmental Research Laboratory, U.S.EPA, Ada, OK.

28 Driscoll, F. G. 1986. *Ground Water and Wells, 2nd Ed.*, Johnson Division, UOP, Inc., St. Paul, MN.

29 Campbell, M. D. and J. H. Lehr. 1973. *Water Well Technology*, McGraw-Hill Book Co., New York, NY.

30 Dobrin, M. B. 1976. *Introduction to Geophysical Prospecting, 3rd Ed.*, McGraw-Hill, New York, NY, p. 630.

31 Keys, W. S. and L. M. MacCary. 1971. "Application of Borehole Geophysics to Water-Resources Investigations," in *Techniques of Water-Resources Investigations,* U.S. Geological Survey, Book 2, Chapter E1.

32 Stewart, M., M. Layton and T. Lizanec. 1983. "Application of Surface Resistivity Surveys to Regional Hydrogeologic Reconnaissance," *Ground Water,* 21:42–48.

33 Kwader, T. 1986. "The Use of Geophysical Logs for Determining Formation Water Quality," *Ground Water,* 24:11–15.

34 Ferris, J. G., D. B. Knowles, R. H. Brown and R. W. Stallman. 1962. "Theory of Aquifer Tests," U.S. Geological Survey Water Supply Paper, 1536-E, pp. 69–174.

35 Kruseman, G. P. and N. A. DeRidder. 1976. "Analysis and Evaluation of Pumping Test Data," International Institute of Land Reclamation and Improvement, Wageningen, The Netherlands, p. 200.

36 Hvorslev, M. J. 1951. "Time Lag and Soil Permeability in Groundwater Observations," *U.S. Army Corps Engrs. Waterways Exp. Sta. Bull.,* 36, Vicksburg, MS.

37 Cooper, H. H., Jr., J. D. Bredehoeft and S. S. Papadopulos. 1967. "Response of a Finite Diameter Well to an Instantaneous Charge of Water," *Water Resourc. Res.,* 3(1):263–269.

38 Papadopulos, I. S., J. D. Bredehoeft and H. H. Cooper, Jr. 1973. "On the Analysis of 'Slug Test' Data," *Water Resourc. Res.,* 9(4):1087–1089.

39 Bouwer, H. and R. C. Rice. 1976. "A Slug Test for Determining Hydraulic Conductivity of Unconfined Aquifers and Completely or Partially Penetrating Wells," *Water Resour. Res.,* 12:423–428.

40 Streltsova, T. D. 1988. *Well Testing in Heterogeneous Formations,* John Wiley & Sons, New York, NY.

41 Snow, D. T. 1968. "Rock Fracture Spacings, Openings, and Porosities," *J. Soil Mech.,* Fourd Div., Proc. ASCE, v. 94, pp. 73–91.

42 Lattman, L. H. and R. R. Parizek. 1964. "Relationship between Fracture Traces and the Occurrence of Ground Water in Carbonate Rocks," *Journal of Hydrology,* 2:73–91.

43 Bogli, A. 1980. *Karst Hydrology and Physical Speleology,* Springer-Verlag, New York, NY.

44 Journel, A. G. and C. J. Huijbregts. 1978. *Mining Geostatistics,* Academic Press, London, England.

45 Englund, E. and A. Sparks. 1988. "GEO-EAS (Geostatistical Environment Assessment Software) User's Guide," U.S. Environmental Protection Agency, EPA/600/4-88/033a, Las Vegas, NV.

46 Cooper, R. M. and J. D. Istok. 1988a. "Geostatistics Applied to Groundwater Pollution. I: Methodology," *Journal of Environmental Engineering,* ASCE, 114(2).

47 Cooper, R. M. and J. D. Istok. 1988b. "Geostatistics Applied to Groundwater Contamination. II: Methodology," *Journal of Environmental Engineering,* ASCE, 114(2).

48 McDonald, M. G. and A. W. Harbaugh. 1984. "A Modular Three-Dimensional Finite-Difference Groundwater Flow Model," U.S. Geological Survey, Open File Report 83–875.

49 Javandel, I., C. Doughty and C. F. Tsang. 1984. *Groundwater Transport: Handbook of Mathematical Models,* American Geophysical Union, Water Resources Monograph 10, Washington, D.C., p. 228.

50 Konikow, L. F. and J. D. Bredehoeft. 1978. "Computer Model of Two-Dimensional Solute Transport and Dispersion in Ground Water, Automated Data Processing and Computations," Techniques of Water Resources Investigations of the U.S. Geological Survey, Washington, D.C.

51 Rifai, H. S., P. B. Bedient, J. T. Wilson, K. M. Miller and J. M. Armstrong. 1988. "Biodegradation Modeling at a Jet Fuel Spill Site," *J. Environment Engr. Div.*, ASCE, 114:1007–1019.

52 Ehrenfeld, J. and J. Bass. 1984. *Evaluation of Remedial Action Unit Operations at Hazardous Waste Disposal Sites*, Noyles Publications, Park Ridge, NJ, p. 434.

53 Canter, L. W. and R. C. Knox. 1986. *Ground Water Pollution Control*, Lewis Publishers, Inc., Chelsea, MI, p. 526.

54 Thomas, J. M., M. D. Lee, P. B. Bedient, R. C. Borden, L. W. Canter and C. H. Ward. 1987. *Leaking Underground Storage Tanks: Remediation with Emphasis on In Situ Biorestoration*, NCGWR, Robert S. Kerr Environmental Research Laboratory, U.S. Environmental Protection Agency, Ada, OK, EPA/600/2-87/008.

55 Mercer, J. W., D. C. Skipp and D. Giffin. 1990. *Basics of Pump-and-Treat Ground-Water Remediation Technology*, Robert S. Kerr Environmental Research Laboratory, U.S. Environmental Protection Agency, Ada, OK, EPA/600/8-90/003.

56 U.S. Environmental Protection Agency. 1989. "Evaluation of Ground Water Extraction Remedies," Vol. 1, Summary Report EPA/540/2-89/054, Washington, D.C.

57 Johnson, P. C., C. C. Stanely, M. W. Kemblowski, D. L. Byers and J. D. Colthart. 1990a. "A Practical Approach to the Design, Operation, and Monitoring of In-Situ Soil-Venting Systems," *Ground Water Monitoring Review*, 10:2:159–178.

58 Johnson, P. C., M. W. Kemblowski and J. D. Colthart. 1990b. "Quantitative Analysis for the Cleanup of Hydrocarbon-Contaminated Soils by In-Situ Venting," *Ground Water*, 28:3:413–429.

59 Javandel, I. and C. F. Tsang. 1986. "Capture-Zone Type Curves: A Tool for Aquifer Cleanup," *Ground Water*, 24(5):616–625.

60 Freeberg, K. M., P. B. Bedient and J. A. Conner. 1987. "Modeling of TCE Contamination and Recovery in a Shallow Sand Aquifer," *Ground Water*, 25:70–80.

61 Konikow, L. F. and J. D. Bredehoeft. 1978. "Computer Model of Two-Dimensional Solute Transport and Dispersion in Ground Water," U.S. Geological Survey Techniques of Water-Resources Investigation, Book 7, Chapter C2, p. 90.

62 Schwarzenback, R. P. and W. Giger. 1985. "Behavior and Fate of Halogenated Hydrocarbons in Ground Water," *Ground Water Quality*, C. H. Ward, W. Giger and P. L. McCarty, eds., New York, Wiley-Interscience, pp. 446–471.

63 Bouwer, E. J., B. E. Rittman and P. L. McCarty. 1981. "Anaerobic Degradation of Halogenated 1- and 2-Carbon Organic Compounds," *Environ. Sci. Technol.*, 15:590.

CHAPTER 6

Bioremediation

INTRODUCTION

ONE of the remediation methods that has been gaining more widespread attention recently is bioremediation, the treatment of subsurface pollutants by stimulating the native microbial population. The concept is to biodegrade complex hydrocarbon pollutants into simple carbon dioxide and water. The technology is not novel; the biodegradation potential of organic contaminants has been recognized and utilized in the wastewater treatment process for years. Suspended growth processes include activated sludge reactors, lagoons, waste stabilization ponds, and fluidized bed reactors. Fixed film processes include trickling filters, rotating biological discs, and sequencing batch reactors (Lee et al., 1988 [1]).

The ultimate goal of the biodegradation process is to convert organic wastes into biomass and CO_2, CH_4, and inorganic salts. Two essential criteria must be in place before biodegradation or bioremediation can occur. First, the subsurface geology must have a relatively large hydraulic conductivity (K) to allow the transport of oxygen and nutrients through the aquifer. Second, microorganisms must be present in sufficient numbers and types to degrade the contaminants of interest. Formations with K values greater than 10^{-4} cm/sec are most amenable to in situ bioremediation.

The natural biodegradation process is simply a biochemical reaction that is mediated by microorganisms found to exist naturally in the subsurface. An organic compound is oxidized (loses hydrogen electrons) by an electron acceptor, which in itself is reduced (gains hydrogen electrons). Several electron acceptors have been identified to date: oxygen (O_2), nitrate (NO_3^-), sulfate (SO_4^{2-}), or carbon dioxide (CO_2). The utilization of oxygen as an electron acceptor is termed aerobic biodegradation and that of nitrate is called anaerobic biodegradation. An example of the aerobic biodegradation reaction for benzene is given by:

$$C_6H_6 + 7.5\ O_2 \rightarrow 6\ CO_2 + 3\ H_2O \qquad (6.1)$$

117

BIODEGRADATION OF ORGANICS IN THE SUBSURFACE

The interest in aerobic biodegradation as a cleanup technology was spurred by early studies such as that presented by Dunlap and McNabb, 1973 [2], which indicated the presence of active microbial populations in the subsurface. The microbial populations were metabolically active, and often nutritionally diverse. The most aerobically biodegradable compounds in the subsurface have been petroleum hydrocarbons such as gasoline, crude oil, heating oil, fuel oil, lube oil waste, and aviation gas (Lee et al., 1987 [3]). In particular, benzene, toluene, xylene, and ethylbenzene are very degradable. Other compounds such as alcohols (isopropanol, methanol, ethanol), ketones (acetone, methyl ethyl ketone), and glycols (ethylene glycol) are also aerobically biodegradable. Recent studies have expanded the list of aerobically degraded compounds to include methylated benzenes, chlorinated benzenes (Kuhn et al., 1985 [4]), chlorinated phenols (Suflita and Miller, 1985 [5]), methylene chloride (Jhaveri and Mazzacca, 1983 [6]), napthalene, methylnapthalenes, dibenzofuran, fluorene, and phenanthrene (Wilson et al., 1985 [7]; Lee and Ward, 1985a [8]).

The sampling methods for subsurface microorganisms have been described in detail by Dunlap et al., 1977 [9] and Wilson et al., 1983 [10]. Basically, the method consists of procuring a soil sample with a core barrel and extruding a sample through a sterile paring device (Lee et al., 1988 [1]). Typical microbial numbers in uncontaminated shallow aquifers range from 10^6 to 10^7 cells/g dry soil, and bacteria is the predominant form of microorganisms observed in the subsurface (Wilson et al., 1983 [10]; Ghiorse and Balkwill, 1985 [11]; White et al., 1983 [12]).

Laboratory studies of biodegradation have been performed to support the successful implementation of the bioremediation technology at the field scale. There are several questions that need to be answered in laboratory experiments. Is a compound biodegradable and under what conditions? What numbers and types of microorganisms are required for degradation? Would the microorganisms respond to stimulation if growth-limiting nutrients are added? At what rates would a given organic chemical degrade?

BIOREMEDIATION PROCESSES

Enhanced aerobic bioremediation is essentially an engineered delivery of nutrients and oxygen, in the form of air, pure oxygen, or hydrogen peroxide, to the contaminated zone of an aquifer.

ADVANTAGES

The main advantage of biological treatment is that it offers partial or complete breakdown of the contaminant instead of simply transferring the contaminant from one phase in the environment to another. Many of the other alternatives for remediation may not have the potential to reduce contaminants to the required levels at a reasonable economic cost. Lee et al., 1988 [1] and Thomas et al., 1987 [13] have discussed in detail the advantages and disadvantages of bioremediation (Table 6.1).

Bioremediation can often be used to treat contaminants that are sorbed to soil or trapped in pore spaces. In addition to treatment of the saturated zone, organics held in the unsaturated and capillary zone can sometimes be treated when an infiltration gallery or soil flushing is used. The time required to treat subsurface pollution using bioremediation can often be faster than some pump-and-treat procedures.

Bioremediation can cost less than other remedial options. Flathman and Githens, 1985 [14] estimated that the cost of bioremediation would be one-fifth of that for excavation and disposal of soil contaminated with isopropanol and tetrahydrofuran and in addition would provide an ultimate

TABLE 6.1. Advantages and Disadvantages of Bioremediation (from Thomas et al., 1987 [13]).

Advantages
(1) Can be used to treat hydrocarbons and certain organic compounds, especially water-soluble pollutants and low levels of other compounds that would be difficult to remove by other methods
(2) Environmentally sound because it does not usually generate waste products and typically results in complete degradation of the contaminants
(3) Utilizes the indigenous microbial flora and does not introduce potentially harmful organisms
(4) Fast, safe, and generally economical
(5) Treatment moves with the groundwater
(6) Good for short-term treatment of organic contaminated groundwater
Disadvantages
(1) Can be inhibited by heavy metals and some organics
(2) Bacteria can plug the soil and reduce circulation
(3) Introduction of nutrients could adversely affect nearby surface waters
(4) Residues may cause taste and odor problems
(5) Labor and maintenance requirements may be high, especially for long-term treatment
(6) May not work for aquifers with low permeabilities that do not permit adequate circulation of nutrients
(7) Long-term effects are unknown

disposal solution. The areal zone of treatment using bioremediation can be larger than other remedial technologies because the treatment moves with the plume and can reach areas which are otherwise inaccessible.

DISADVANTAGES

There are also disadvantages to bioremediation. Many organic compounds in the subsurface are resistant to degradation. Bioremediation requires an acclimated population; however, adapted populations may not develop for recent spills or recalcitrant compounds. Heavy metals and toxic concentrations of organics may inhibit microbial activity and preclude the use of the indigenous microorganisms for bioremediation at some sites.

The pumping and injection wells may clog from excessive microbial growth which results from the addition of oxygen and nutrients. In one biostimulation project, microbial growth produced foaming in the well casings (Raymond et al., 1976 [15]). In addition, the hydrodynamics of the restoration program must be properly managed. The nutrients added must be contained within the treatment zone to prevent eutrophication of untargeted areas. High concentrations of nitrate can render groundwater unpotable. Metabolites of partial degradation of organic compounds may impart objectionable tastes and odors.

Biostimulation projects require extensive and continuous monitoring and maintenance for successful treatment; whether these requirements are greater than those for other remedial actions is debatable. The process results in increased microbial biomass that could decompose and release undesirable metabolites. In addition, microbial growth can exert an oxygen demand that may drive the system anaerobic and result in the production of hydrogen sulfide or other objectionable byproducts.

There are other factors that may limit or inhibit the biodegradation of subsurface organic pollutants, even in the presence of microorganisms. These factors include the concentration of the contaminant, the presence of toxicants, sorption, pH, and temperature (Lee et al., 1988 [1]). The lack of oxygen and other inorganic nutrients such as phosphorus and nitrogen has been extensively documented by several researchers to be limiting factor to biodegradation in laboratory and field scale experiments (Alexander, 1980 [16]; Lee and Ward, 1985a [8]). At the field scale, the transport characteristics of the aquifer limit the delivery of the required nutrients to the microbial population. By increasing the rates at which required nutrients are delivered to microorganisms, essentially the main concept behind bioremediation, one can enhance the natural biodegradation process and actually achieve faster removal rates.

DESIGN OF A BIOREMEDIATION PROCESS

The basic steps involved in an in situ bioremediation program (Lee et al., 1988 [1]) are: (1) site investigation, (2) free product recovery, (3) microbial degradation study, (4) system design, (5) operation, and (6) maintenance. As in pump-and-treat systems, it is important to define the hydrogeology and the extent of contamination at the site prior to the initiation of any in situ effort. The parameters of interest at a field site include the direction and rate of groundwater flow, the depths to the water table and to the contaminated zone, the specific yield of the aquifer, and the heterogeneity of the soil. In addition, other parameters such as hydraulic connections between aquifers, potential recharge and discharge zones, and seasonal fluctuations of the water table should be considered.

INVESTIGATION AND FEASIBILITY STUDIES

A number of monitoring wells should be installed and slug tests and pump tests should be performed for determining hydraulic conductivity. Water levels should be measured in the wells to determine a potentiometric map of the site. Monitoring wells should be sampled for presence of hydrocarbon contamination and a plume should be mapped for the site. The pumping rate that can be sustained in the aquifer is an important consideration because it limits the amount of water that can be circulated in the system during the bioremediation process. Hydraulic conductivity should exceed about 10^{-4} cm/sec for successful bioremediation.

After defining the hydrogeology, recovery of any free product at the site should be completed (see Chapter 8). For fuels less dense than water, the nonaqueous phase liquid (NAPL) can be removed using physical recovery techniques such as a single pump system that produces water and hydrocarbon or a two-pump, two-well system that steepens the hydraulic gradient and recovers the accumulating hydrocarbon.

Prior to the initiation of a bioremediation activity, it is important to conduct a feasibility study for the biodegradation of the contaminants present at the site. First, contaminant-degrading microorganisms must be present, and second, the response of these native microorganisms to the proposed treatment method must be evaluated. In addition, the feasibility study is conducted to determine the nutrient requirements of the microorganisms. These laboratory studies provide a reliable basis for performance at the field level only if they are performed under conditions that simulate the field.

Feasibility studies can be completed using several different techniques. Batch culture techniques are used to measure the disappearance of the contaminant; electrolytic respirometer studies are utilized to measure the up-take of oxygen. Tests which are designed to measure an increase in micro-bial numbers are not sufficient indicators of metabolization of the contaminant in question. Instead, studies that measure disappearance of the contaminant or mineralization studies that confirm the breakdown of the contaminant to carbon dioxide and water need to be conducted. Con-trols to detect abiotic transformation of the pollutants and tests to detect toxic effects of the contaminants on the microflora should be included (Flathman et al., 1984 [17]).

HYDRAULIC REQUIREMENTS FOR BIOREMEDIATION

A system for injection of nutrients into the formation and circulation through the contaminated portion of the aquifer must be designed and con-structed (Lee and Ward, 1985b [18]). The system usually includes injection and production wells and equipment for the addition and mixing of the nutrient solution (Raymond, 1978 [21]). A typical system is shown in Figure 6.1. Placement of injection and production wells may be restricted by the presence of physical structures. Wells should be screened to accom-modate seasonal fluctuations in the level of the water table. Wells are often installed within the contaminant plume to monitor concentrations of hydrocarbon, nutrients, microbes, and other parameters. Monitoring of the process is critical for evaluating its success or failure. Microbial cell counts and contaminant breakthrough curves are indicators of how well the process is performing.

Well installation should be performed under the direction of a hydro-geologist to ensure adequate circulation of the groundwater (Lee and Ward, 1985b [18]). Produced water can be recycled to recirculate unused nutrients, avoid disposal of potentially contaminated groundwater, and avoid the need for makeup water. Inorganic nutrients can be added to the subsurface once the system is constructed. Continuous injection of the nutrient solution is labor-intensive but provides a more constant nutrient supply than a discontinuous process. Continuous addition of oxygen is recommended because the oxygen is likely to be a limiting factor in hydrocarbon degradation.

The rate of aerobic biodegradation is usually limited by the amount of oxygen that can be transported to the organisms in the zone of contamina-tion. A number of methods are available to supply oxygen to groundwater, including the addition of air, pure oxygen, or hydrogen peroxide, with in-creasing concentrations, respectively. Sparging the groundwater with air and pure oxygen can supply only 8 to 40 mg/L of oxygen, depending on the temperature of the injection fluid (Lee et al., 1988 [1]).

Addition of hydrogen peroxide must be carefully monitored to avoid

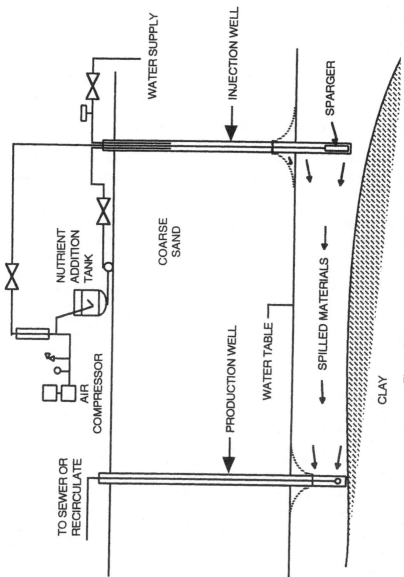

Figure 6.1 Injection system for oxygen.

123

being toxic to microorganisms at elevated concentrations (Thomas and Ward, 1989 [20]). Air can be supplied with carborundum diffusers (Raymond et al., 1975 [19]), by smaller diffusers constructed from a short piece of DuPont Viaflo tubing (Raymond, 1978 [21]), or by diffusers spaced along air lines buried in the injection lines (Minugh et al., 1983 [22]). The size of the compressor and the number of diffusers are determined by the extent of contamination and the time allowed for treatment (Raymond, 1978 [21]).

CHEMICAL AND BIOLOGICAL REQUIREMENTS

Nutrients also can be circulated using an infiltration gallery (Figure 6.2); this method provides an additional advantage of treating the residual gasoline that may be trapped in the pore spaces of the unsaturated zone (Brenoel and Brown, 1985 [23]). Oxygen also can be supplied using hydrogen peroxide, ozone, or soil venting. The performance of the system and proper distribution of the nutrients can be monitored by measuring the organic, inorganic, and bacterial levels (Lee and Ward, 1985b [18]).

Bioremediation is not without its problems, however. The most important is the lack of well-documented field demonstrations that show the effectiveness of the technology and what if any are the long-term effects of this treatment on groundwater systems. Other problems include the possibility for generating undesirable intermediate compounds during the biodegradation process which are more persistent in the environment than the parent compound.

Hydrogen peroxide, which dissociates to form water and oxygen, is infinitely soluble in water (Thomas and Ward, 1989 [20]); however, hydrogen peroxide can be toxic to microorganisms at concentrations as low as 100 ppm. A stepping-up procedure is usually utilized to allow the microorganisms to adapt to the higher concentrations of the oxidant. Other problems have to do with the stability of hydrogen peroxide. The key to success in using hydrogen peroxide as an oxygen source is to add a relatively large quantity to water and have oxygen released in a controlled manner as it advances through the aquifer. If hydrogen peroxide is destabilized, oxygen will come out of solution as a gas, and the process becomes less efficient (Hinchee et al., 1987 [24]). Proprietary techniques have been developed to stabilize hydrogen peroxide.

In severe cases, gas production (both O_2 and CO_2) can lead to a reduction in hydraulic conductivity (Brown et al., 1984 [25]). One undesirable effect of using hydrogen peroxide is that other redox reactions may also be enhanced. A rather important drawback for the use of hydrogen peroxide may be its cost; at 1987 prices, the cost of 35% H_2O_2 was \$4.20 per gallon. Depending on the size of the spill and the required amount of oxygen, the cost of materials could make bioremediation an expensive alternative.

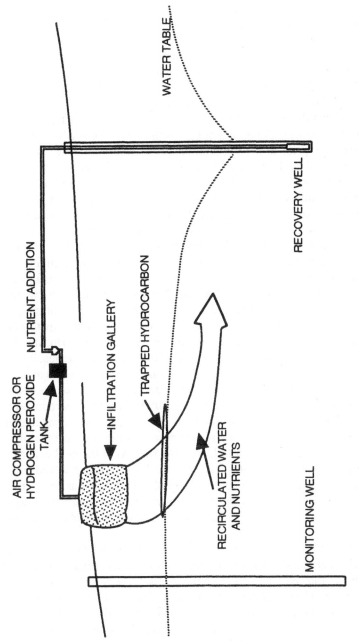

Figure 6.2 Infiltration gallery.

125

EXAMPLES AND DETAILED TABULATION OF CASE STUDIES

Researchers from Suntech, Inc. are amongst the earliest pioneers who utilized bioremediation at sites contaminated with gasoline. Two field experiments are discussed by Jamison et al., 1975 [26]; Raymond et al., 1977 [27]; Raymond et al., 1978 [28]. The first field study was at a site in Ambler, Pennsylvania. A leak in a gasoline pipeline had caused the township to abandon its groundwater supply wells. The free product was physically removed prior to the initiation of biodegradation studies at the site. Laboratory studies showed that the natural microbial population at the site could use the spilled high-octane gasoline as the sole carbon source if sufficient quantities of the limiting nutrients, in this case, oxygen, nitrogen, and phosphate, were supplied. Pilot studies that were carried out in the field in several wells confirmed the laboratory findings.

The second field experiment completed by the Suntech, Inc. group was at a site in Millville, New Jersey. Again laboratory studies utilizing sands from the aquifer confirmed the ability of the natural microorganisms to degrade the spilled gasoline at the site. The main difference between the Ambler site and the Millville site was the hydrogeologic nature of the media. The aquifer at the Millville site consisted of medium to coarse well-sorted quartz sand while that at Ambler was a highly permeable limestone aquifer. The results from the field effort supported the feasibility of bioremediation at sites with consolidated and unconsolidated sands.

In a controlled field experiment at the Canada Forces Base Borden site, two plumes of gasoline-contaminated groundwater were introduced into the aquifer. Immediately upgradient of one plume, groundwater spiked with nitrate was added so that a nitrate plume would overtake the organic plume (Berry-Spark et al., 1986 [29]). The success of the field experiment was limited (Berry-Spark and Barker, 1987 [30]). The dissolved organic contaminant mass (BTEX) decreased rapidly due to residual oxygen concentrations in the aquifer prior to the nitrate overlap. Insufficient organic mass left in the aquifer was not adequate to evaluate anaerobic biotransformation.

Semprini et al., 1988 [31] presented the results from a field evaluation of in situ biodegradation of trichloroethylene (TCE) and related compounds. The method that was used in the field demonstration relied on the ability of methane-oxidizing bacteria to degrade these contaminants to stable, nontoxic end products. The field site is located at the Moffett Naval Air Station in Mountain View, California and the test zone is a shallow confined aquifer composed of coarse-grained alluvial sediments. Results from the biotransformation experiments at the site indicate that biodegradation of TCE was on the order of 30% of the mass injected.

Major et al., 1988 [32] investigated the biodegradation of benzene, toluene, and the isomers of xylene (BTX) in anaerobic batch microcosms containing shallow aquifer material. BTX loss occurred with the addition of either nitrate or oxygen. Denitrification was confirmed by nitrous oxide accumulation. When a limiting amount of nitrate was added, there was a corresponding limit to the loss of BTX and a limited amount of nitrous oxide production.

Chiang et al., 1989 [33] characterized soluble hydrocarbon and dissolved oxygen in a shallow aquifer beneath a field site by sampling groundwater at forty-two monitoring wells. Results from ten sampling periods over three years showed a significant reduction in total benzene mass with time in groundwater. The natural attenuation rate was calculated to be 0.95% per day. Spatial relationships between DO and total benzene, toluene, and xylene (BTX) were shown to be strongly correlated by statistical analyses and solute transport modeling.

UNITED CREOSOTE CO. SUPERFUND SITE

Borden and Bedient, 1986 [34]; Borden et al., 1986 [35] along with the EPA RS Kerr Environmental Research Lab evaluated the natural biodegradation occurring at an abandoned United Creosote Company (UCC) site in Conroe, Texas over a four-year period. They observed significant reductions of naphthalene in the contaminant plume compared to a conservative chloride tracer (see Figure 6.3). The authors modeled the biodegradation by applying the concept of instantaneous reaction of hydrocarbon with oxygen, which evolved into the BIOPLUME model described below.

Borden and Bedient, 1987 [36] conducted a three-well injection-production test at the UCC site in Conroe, Texas, to evaluate the significance of biotransformation in limiting the transport of polycyclic aromatics present in the shallow aquifer. During the test, chloride, a nonreactive tracer and two organic compounds, naphthalene and para-dichlorobenzene (pDCB), were injected into a center well for twenty-four hours followed by clean groundwater for six days. Groundwater was continuously produced from two adjoining wells and monitored to observe the breakthrough of these compounds (Figure 6.4).

Data were statistically analyzed using method of moments analyses between the chloride and organics curves. Retardation was not very significant with factors ranging from 0.97 to 1.03. Figure 6.5 indicates a significant loss of naphthalene, and pDCB attributed to biotransformation processes was observed during the test. These results suggested that biodegradation is the major process limiting the transport of naphthalene and other similar compounds at the site.

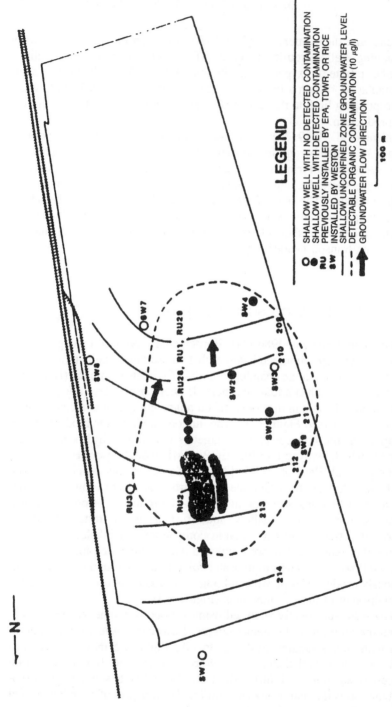

LEGEND

○ SHALLOW WELL WITH NO DETECTED CONTAMINATION
● SHALLOW WELL WITH DETECTED CONTAMINATION
RU PREVIOUSLY INSTALLED BY EPA, TDWR, OR RICE
SW INSTALLED BY WESTON
— SHALLOW UNCONFINED ZONE GROUNDWATER LEVEL
╌╌ DETECTABLE ORGANIC CONTAMINATION (10 μg/l)
➤ GROUNDWATER FLOW DIRECTION

100 m

Figure 6.3 United Creosote Co., Inc. site map (from Borden and Bedient, Water Resources Bulletin, v. 23, p. 630, 1987, copyright by American Water Resources Association).

128

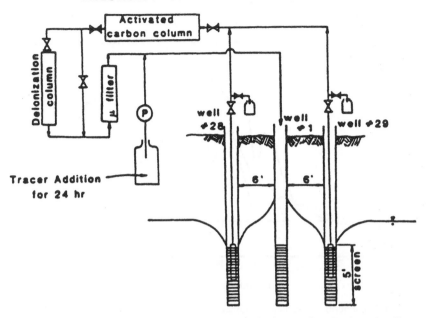

Figure 6.4 Treatment system used during the injection-production test (from Borden and Bedient, Water Resources Bulletin, v. 23, p. 632, 1987, copyright by American Water Resources Association).

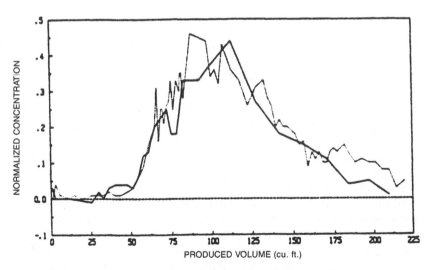

Figure 6.5 Comparison of normalized concentration distributions for chloride (———) and conductivity (·······) in well RU 29 (from Borden and Bedient, Water Resources Bulletin, v. 23, p. 634, 1987, copyright by American Water Resources Association).

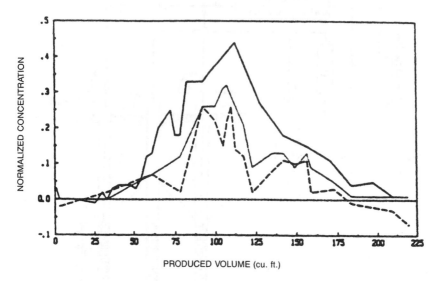

PRODUCED VOLUME (cu. ft.)

Figure 6.5 (continued) Comparison of normalized concentration distributions for chloride (————), pDCB (·······), and naphthalene (--------) in RU 29 (from Borden and Bedient, Water Resources Bulletin, v. 23, p. 634, 1987, copyright by American Water Resources Association).

AVIATION FUEL SPILL AT TRAVERSE CITY, MICHIGAN

Rifai et al., 1988 [37] studied the naturally occurring biodegradation at an aviation fuel spill at the U.S. Coast Guard Station in Traverse City, Michigan (Figure 6.6). Contamination data from approximately twenty-five wells at the site were utilized to define the dissolved benzene, toluene, and xylene (BTEX) plume over a two-year period (Figure 6.7). The plume was originally over 5000 ft in length and several hundred ft wide, contaminating a nearby subdivision of homes using shallow groundwater. An interdiction field of wells was installed at the site boundary to limit the problem to the base (Figure 6.6). The site has become one of the best monitored and studied leaking underground tank sites in the U.S.

Based on a careful study of BTEX plume data from twenty-five wells vertically averaged over a two-year period, Rifai et al., 1988 [37] calculated a biodegradation rate of 1.0% a day at the Traverse City site. Figure 6.8 indicates that the pump-and-treat system at the site was removing a very small percentage of the total dissolved mass present at the site. A modeling effort was completed using the BIOPLUME II model. The modeling results along the centerline of the contaminant plume were good, and the BIOPLUME II model results matched the field observations except in an area between monitoring wells M30 and the pumping wells.

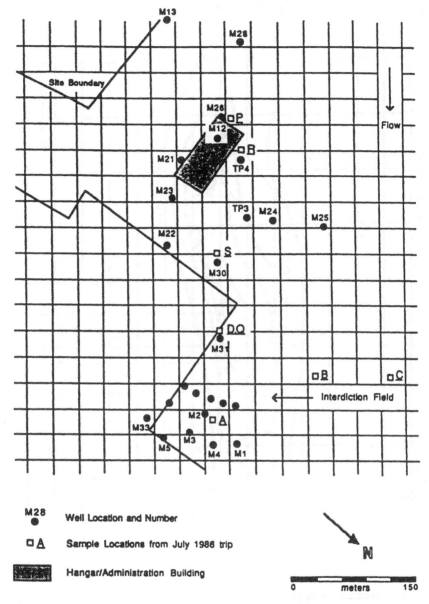

Figure 6.6 Traverse City field site map (from Rifai et al., *Journal of Environmental Engineering*, v. 114, p. 1015, 1988, copyright by ASCE).

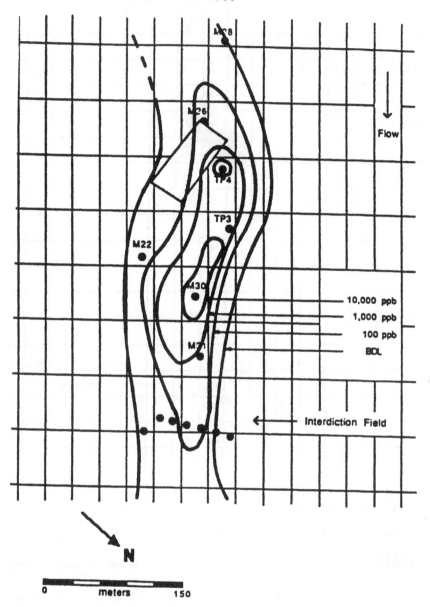

Figure 6.7 Traverse City field site—BTX plume (quarter 2, 1986) (from Rifai et al., *Journal of Environmental Engineering*, v. 114, p. 1021, 1988, copyright by ASCE).

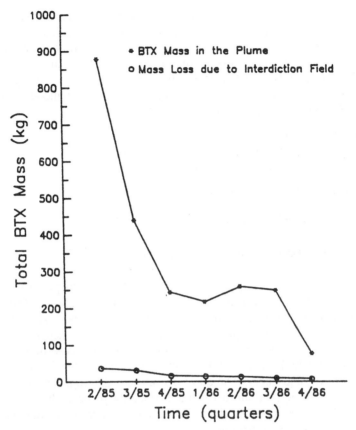

Figure 6.8 Variation in total BTX with time at Traverse City field site (from Rifai et al., *Journal of Environmental Engineering*, v. 114, p. 1024, 1988, copyright by ASCE).

The site has also been the location of a major effort to test and quantify enhanced in situ bioremediation using oxygen, hydrogen peroxide, and nitrate injections in wells and galleries. The tests were extensive in that vertical and horizontal sampling were conducted for organics and dissolved oxygen over a long period of time followed by actual core sampling near the water table. The results of the tests were relatively successful in biodegrading the soluble portions of the BTEX plume into which oxygen and nutrients were injected.

APPLICATION OF MODELS TO BIODEGRADATION PROCESSES

The problem of quantifying biodegradation in the subsurface can be addressed by using models that combine physical, chemical, and biological processes. Developing such models is not simple, however, due to the complex nature of microbial kinetics, the limitations of computer resources, the lack of field data on biodegradation, and the lack of robust numerical schemes that can simulate the physical, chemical, and biological processes accurately. Several researchers have developed groundwater biodegradation models. The main approaches utilized for modeling biodegradation kinetics are:

(1) First-order decay
(2) Biofilm models (include kinetic expressions)
(3) Instantaneous reaction model
(4) Dual-substrate Monod model

These are described in more detail in the next section.

REVIEW OF PREVIOUS MODELING EFFORTS

McCarty et al., 1981 [38] modeled the biodegradation process using biofilm kinetics. They assumed that substrate concentration within the biofilm changes only in the direction that is normal to the surface of the biofilm and that all the required nutrients are in excess except the rate-limiting substrate. The model employs three basic processes: mass transport from the bulk liquid, biodecomposition within the biofilm, and biofilm growth and decay. The authors evaluated the applicability of the biofilm model to aerobic subsurface biodegradation using a laboratory column filled with glass beads. The experimental data and the model predictions were relatively consistent.

Kissel et al., 1984 [39] developed differential equations describing mass balances on solutes and mass fractions in a mixed-culture biological film within a completely mixed reactor. The model incorporates external mass transport effects, Monod kinetics with internal determination of limiting electron donor or acceptor, competitive and sequential reactions, and multiple active and inert biological fractions that vary spatially. Results of hypothetical simulations involving competition between heterotrophs deriving energy from an organic solute and autotrophs deriving energy from ammonia and nitrite were presented.

Molz et al., 1986 [40] and Widdowson et al., 1987 [41] presented one-dimensional and two-dimensional models for aerobic biodegradation of organic contaminants in groundwater coupled with advective and dispersive transport. A microcolony approach was utilized in the modeling effort. Microcolonies of bacteria are represented as disks of uniform radius and thickness attached to aquifer sediments. A boundary layer of a given thickness was associated with each colony across which substrate and oxygen are transported by diffusion to the colonies. Their results indicated that biodegradation would be expected to have a major effect on contaminant transport when proper conditions for growth exist. Simulations of two-dimensional transport suggested that under aerobic conditions, microbial degradation reduces the substrate concentration profile along longitudinal sections of the plume and retards the lateral spread of the plume. Anaerobic conditions developed in the plume center due to microbial consumption and limited oxygen diffusion into the plume interior.

Widdowson et al., 1988 [42] also extended their previous work to simulate oxygen- and/or nitrate-based respiration. Basic assumptions incorporated into the model include a simulated particle-bound microbial population comprised of heterotrophic, facultative bacteria in which metabolism is controlled by lack of either an organic carbon-electron donor source (substrate), electron acceptor (O_2 and/or NO_3), or mineral nutrient (NH_4^+), or all three simultaneously.

Srinivasan and Mercer, 1988 [43] presented a one-dimensional, finite difference model for simulating biodegradation and sorption processes in saturated porous media. The model formulation allows for accommodating a variety of boundary conditions and process theories. Aerobic biodegradation was modeled using a modified Monod function; anaerobic biodegradation is modeled using Michaelis-Menten kinetics. In addition, first-order degradation was allowed for both substances. Sorption was incorporated using linear, Freundlich, or Langmuir equilibrium isotherms for either substance.

MacQuarrie et al., 1990 [44] utilized the model developed by MacQuarrie and Sudicky, 1990 [45] to examine plume behavior in uniform and random flow fields. In uniform groundwater flow, a plume originating from a high-concentration source will experience more spreading and slower normalized mass loss than a plume from a lower initial concentration source because dissolved oxygen is more quickly depleted. Large groundwater velocities produced increases in the rate of organic solute mass loss because of increased mechanical mixing of the organic plume with oxygenated groundwater.

DEVELOPMENT AND APPLICATION OF BIOPLUME

Borden and Bedient, 1986 [34] developed the first version of the BIOPLUME model. They developed a system of equations to simulate the simultaneous growth, decay, and transport of microorganisms combined with the transport and removal of hydrocarbons and oxygen. Simulation results indicated that any available oxygen in the region near the hydrocarbon source will be rapidly consumed. In the body of the hydrocarbon plume, oxygen transport will be rate limiting and the consumption of oxygen and hydrocarbon can be approximated as an instantaneous reaction. The major sources of oxygen, the researchers concluded, are transverse mixing, advective fluxes, and vertical exchange with the unsaturated zone.

Borden and Bedient, 1986 [34]; Borden et al., 1986 [35] applied the first version of the BIOPLUME model to simulate biodegradation at the Conroe Superfund site in Texas. Oxygen exchange within the unsaturated zone was simulated as a first-order decay in hydrocarbon concentration. The loss of hydrocarbon due to horizontal mixing with oxygenated ground-water and resulting biodegradation was simulated by generating oxygen and hydrocarbon distributions independently and then combining by superposition (Figure 6.9). Simulated oxygen and hydrocarbon concentrations closely matched the observed values at the Conroe site.

Rifai et al., 1987a [46], 1987b [47], 1988 [37] expanded and extended the original BIOPLUME and developed a numerical version of the biodegradation model (BIOPLUME II) by modifying the USGS two-dimensional

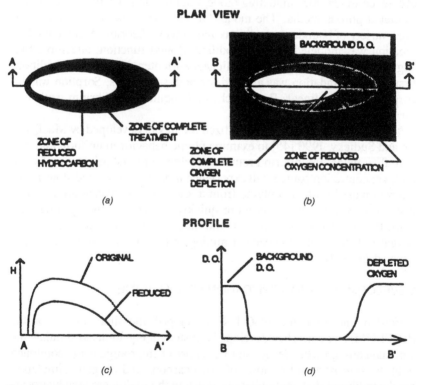

Figure 6.9 Use of principle of superposition for hydrocarbon and oxygen in BIOPLUME II: (a) hydrocarbon plume; (b) oxygen plume; (c) hydrocarbon; (d) oxygen (from Rifai et al., *Journal of Environmental Engineering*, v. 114, p. 1010, 1988, copyright by ASCE).

solute transport model (Konikow and Bredehoeft, 1978 [48]). The basic concept applied in developing BIOPLUME II includes the use of a dual-particle mover procedure to simulate the transport of oxygen and contaminants in the subsurface.

Biodegradation of the contaminants is approximated by the instantaneous reaction model. The ratio of oxygen to dissolved contaminants consumed by the reaction is determined from an appropriate stoichiometric model (assuming complete mineralization). In general, the transport equation is solved twice at every time step to calculate the oxygen and contaminant distributions.

$$\frac{\partial(Cb)}{\partial t} = \frac{1}{R_c}\left[\frac{\partial}{\partial x_i}\left(bD_{ij}\frac{\partial C}{\partial x_j}\right) - \frac{\partial}{\partial x_i}bCV_i)\right] - \frac{C'W}{n} \quad (6.2)$$

$$\frac{\partial(Ob)}{\partial t} = \left[\frac{\partial}{\partial x_i}\left(bD_{ij}\frac{\partial O}{\partial x_j}\right) - \frac{\partial}{\partial x_i}(bOV_i)\right] - \frac{O'W}{n} \quad (6.3)$$

where C and O = concentration of contaminant and oxygen respectively, C' and O' = concentration of contaminant and oxygen in a source or sink fluid, n = effective porosity, b = saturated thickness, t = time, x_i and x_j = cartesian coordinates, W = volume flux per unit area, V_i = seepage velocity in the direction of x_i, R_c = retardation factor for contaminant, and D_{ij} = coefficient of hydrodynamic dispersion.

It is emphasized that the BIOPLUME II model simulates dissolved contaminant concentrations that are vertically averaged over the thickness of the aquifer. The two plumes are combined using the principle of superposition (Figure 6.9) to simulate the instantaneous reaction between oxygen and the contaminants, and the decrease in contaminant and oxygen concentrations is calculated from:

$$\Delta C_{RC} = O/F \quad O = 0 \text{ where } C > O/F \quad (6.4)$$

$$\Delta C_{RO} = C \cdot F \quad C = 0 \text{ where } O > C \cdot F \quad (6.5)$$

where ΔC_{RC} and ΔC_{RO} are the calculated changes in concentrations of contaminant and oxygen, respectively, due to biodegradation.

The only model input parameters to BIOPLUME II that are required to simulate biodegradation include the amount of dissolved oxygen in the aquifer prior to contamination and the oxygen demand of the contaminant determined from a stoichiometric relationship. Other parameters are the same as would be required to run the standard USGS MOC model in two dimensions.

REFERENCES

1 Lee, M. D., J. M. Thomas, R. C. Borden, P. B. Bedient, C. H. Ward and J. T. Wilson. 1988. "Biorestoration of Aquifers Contaminated with Organic Compounds," *NCGWR*, R. S. Kerr Environmental Research Laboratory, U.S. Environmental Protection Agency, Ada, OK, 18(1):29–89.

2 Dunlap, W. J. and J. F. McNabb. 1973. "Subsurface Biological Activity in Relation to Ground Water Pollution," EPA/660/2-73/014, U.S. Environmental Protection Agency, Ada, OK, p. 60.

3 Lee, M. D., V. W. Jamison and R. L. Raymond. 1987. "Applicability of In Situ Bioreclamation as a Remedial Action Alternative," *Proc. of Petroleum Hydrocarbons and Organic Chemicals in Ground Water: Prevention, Detection and Restoration, Houston, Texas*, National Water Well Association, pp. 167–185.

4 Kuhn, E. P., P. J. Colberg, J. L. Schnoor, O. Warner, A. J. B. Zehnder and R. P. Schwarzenbach. 1985. "Microbial Transformation of Substituted Benzenes during Infiltration of River Water to Ground Water: Laboratory Column Studies," *Environ. Sci. Technol.*, 19:961.

5 Suflita, J. M. and G. D. Miller. 1985. "Microbial Metabolism of Chlorophenolic Compounds in Ground Water Aquifers," *Environ. Toxicol. Chem.*, 4:751.

6 Jhaveri, V. and A. J. Mazzacca. 1983. "Bio-Reclamation of Ground and Groundwater. A Case History," in *Proc., 4th Natl. Conf. on Management of Uncontrolled Hazardous Waste Sites, Washington, D.C.*, p. 242.

7 Wilson, J. T., J. F. McNabb, J. W. Cochran, T. H. Wang, M. B. Tomson and P. B. Bedient. 1985. "Influence of Microbial Adaptation on the Fate of Organic Pollutants in Ground Water," *Environ. Toxicol. Chem.*, 4:721.

8 Lee, M. D. and C. H. Ward. 1985a. "Microbial Ecology of a Hazardous Waste Disposal Site; Enhancement of Biodegradation," in *Proc., 2nd Conf. on Groundwater Res. Quality Res.*, N. N. Durham and A. E. Redelfs, eds., Oklahoma State University Printing Services, Stillwater, OK, p. 25.

9 Dunlap, W. J., J. F. McNabb, M. R. Scalf and R. L. Cosby. 1977. "Sampling for Organic Chemicals and Microorganisms in the Subsurface," EPA/600/2-77/176, U.S. Environmental Protection Agency, Ada, OK.

10 Wilson, J. T., J. F. McNabb, D. L. Balkwill and W. C. Ghiorse. 1983. "Enumeration and Characterization of Bacteria Indigenous to a Shallow Water-Table Aquifer," *Ground Water*, 21:134.

11 Ghiorse, W. C. and D. L. Balkwill. 1985. "Microbial Characterization of Subsurface Environments," *Ground Water Quality*, C. H. Ward, W. Giger and P. L. McCarty, eds., John Wiley & Sons, New York, p. 387.

12 White, D. C., G. A. Smith, M. J. Gehron, J. H. Parker, R. H. Findlay, R. F. Martz and H. L. Fredrickson. 1983. "The Ground Water Aquifer Microbiotia: Biomass, Community Structure, and Nutritional Status," *Dev. Ind. Microbiol.*, 24:204.

13 Thomas, J. M., M. D. Lee, P. B. Bedient, R. C. Borden, L. W. Canter and C. H. Ward. 1987. "Leaking Underground Storage Tanks: Remediation with Emphasis on In Situ Biorestoration," *NCGWR*, Robert S. Kerr Environmental Research Laboratory, U.S. Environmental Protection Agency, Ada, OK, EPA/600/2-87/008.

14 Flathman, P. E. and G. D. Githens. 1985. "In Situ Biological Treatment of Iso-

propanol, Acetone, and Tetrahydrofuran in the Soil/Groundwater Environment," *Groundwater Treatment Technology*, E. K. Nyer, ed., Van Nostrand Reinhold, New York, p. 173.

15 Raymond, R. L., V. W. Jamison and J. O. Hudson. 1976. "Beneficial Stimulation of Bacterial Activity in Groundwaters Containing Petroleum Products," *AIChE Symp. Ser.*, 73:390.

16 Alexander, M. 1980. "Biodegradation of Chemicals of Environmental Concern," *Science*, 211:132.

17 Flathman, P. E., J. R. Quince and L. S. Bottomley. 1984. "Biological Treatment of Ethylene Glycol-Contaminated Ground Water at Naval Engineering Center in Lakehurst, New Jersey," in *Proc. 4th Nat. Symp. on Aquifer Restoration and Ground Water Monitoring, Columbus, Ohio, May 1984*, D. M. Nielsen and M. Curl, eds., National Water Well Association, Worthington, Ohio, p. 111.

18 Lee, M. D. and C. H. Ward. 1985b. "Biological Methods for the Restoration of Contaminated Aquifers," *J. Environ. Toxicol. Chem.*, 4:743.

19 Raymond, R. L., V. W. Jamison and J. O. Hudson. 1975. "Final Report on Beneficial Stimulation of Bacterial Activity in Ground Water Containing Petroleum Products," Committee on Environmental Affairs, American Petroleum Institute, Washington, D.C.

20 Thomas, J. M. and C. H. Ward. 1989. "In Situ Biorestoration of Organic Contaminants in the Subsurface," *Environ. Sci. Technol.*, 23(7):760–766.

21 Raymond, R. L. 1978. "Environmental Bioreclamation," *Mid-Continent Conf. and Exhibition on Control of Chemicals and Oil Spills*, Detroit, MI, September.

22 Minugh, E. M., J. J. Patry, D. A. Keech and W. R. Leek. 1983. "A Case History: Cleanup of a Subsurface Leak of Refined Product," in *Proc., Oil Spill Conf.—Prevention, Behavior, Control, and Cleanup, San Antonio, TX*, March 1983, p. 397.

23 Brenoel, M. and R. A. Brown. 1985. "Remediation of a Leaking Underground Storage Tank with Enhanced Bioreclamation," in *Proc., Nat. Symp. Exp. on Aquifer Restoration and Ground Water Monitoring*, NWWA, Worthington, OH, pp. 527.

24 Hinchee, R. E., D. C. Downey and E. J. Coleman. 1987. "Enhanced Bioremediation, Soil Venting, and Ground-Water Extraction; A Cost-Effectiveness and Feasibility Comparison," in *Proc., Petroleum Hydrocarbons and Organic Chemicals in Ground Water*, NWWA, Dublin, OH, pp. 147–164.

25 Brown, R. A., R. D. Norris and R. L. Raymond. 1984. "Oxygen Transport in Contaminated Aquifers," in *Proc., Conference on Petroleum Hydrocarbons and Organic Chemicals in Ground Water*, NWWA, Worthington, OH, pp. 421.

26 Jamison, V. W., R. L. Raymond and J. O. Hudson, Jr. 1975. "Biodegradation of High-Octane Gasoline in Groundwater," *Dev. Ind. Microbiol.*, 16:305.

27 Raymond, R. L., J. O. Hudson and V. W. Jamison. 1977. "Final Report, Bacterial Growth in and Penetration of Consolidated and Unconsolidated Sands Containing Gasoline," American Petroleum Institute, Project No. 307-76, Washington, D.C.

28 Raymond, R. L., V. W. Jamison, J. O. Hudson, R. E. Mitchell and V. E. Farmer. 1978. "Final Report, Field Application of Subsurface Biodegradation of Gasoline in a Sand Formation," American Petroleum Institute Project No. 307-77, Washington, D.C.

29 Berry-Spark, K., J. F. Barker, D. Major and C. I. Mayfield. 1986. "Remediation

of Gasoline-Contaminated Ground-Waters: A Controlled Field Experiment," in *Proc., Petroleum Hydrocarbons and Organic Chemicals in Ground Water*, NWWA, pp. 613–623.

30 Berry-Spark, K. and J. F. Barker. 1987. "Nitrate Remediation of Gasoline Contaminated Ground Waters: Results of a Controlled Field Experiment," in *Proc., Petroleum Hydrocarbons and Organic Chemicals in Ground Water*, NWWA/API, pp. 3–10.

31 Semprini, L. et al. 1988. "A Field Evaluation of In Situ Biodegradation for Aquifer Restoration," U. S. Environmental Protection Agency, U. S. Government Printing Office, Ada, OK, EPA/600/S2-87/096.

32 Major, D. W., C. I. Mayfield and J. F. Barker. 1988. *Ground Water*, 26:8–14.

33 Chiang, C. Y., J. P. Salanitro, E. Y. Chai, J. D. Colthart and C. L. Klein. 1989. "Aerobic Biodegradation of Benzene, Xylene in a Sandy Aquifer—Data Analysis and Computer Modeling," *Ground Water*, 27(6):823–834.

34 Borden, R. C. and P. B. Bedient. 1986. "Transport of Dissolved Hydrocarbons Influenced by Reaeration and Oxygen Limited Biodegradation: 1. Theoretical Development," *Water Resour. Res.*, 22:1973–1982.

35 Borden, R. C., P. B. Bedient, M. D. Lee, C. H. Ward and J. T. Wilson. 1986. "Transport of Dissolved Hydrocarbons Influenced by Oxygen Limited Biodegradation: 2. Field Application," *Water Resour. Res.*, 22:1983–1990.

36 Borden, R. C. and P. B. Bedient. 1987. "In Situ Measurement of Adsorption and Biotransformation at a Hazardous Waste Site," *Water Resour. Bull.*, 23:629–636.

37 Rifai, H. S., P. B. Bedient, J. T. Wilson, K. M. Miller and J. M. Armstrong. 1988. "Biodegradation Modeling at a Jet Fuel Spill Site," *J. Environmental Engr. Div.*, ASCE, 114:1007–1019.

38 McCarty, P. L., M. Reinhard, B. E. Rittmann. 1981. "Trace Organics in Groundwater," *Environ. Sci. Technol.*, 15(1):40–51.

39 Kissel, J. C., P. L. McCarty and R. L. Street. 1984. "Numerical Simulation of Mixed-Culture Biofilm," *J. Environ. Eng. Div.*, ASCE, 110:393.

40 Molz, F. J., M. A. Widdowson and L. D. Benefield. 1986. "Simulation of Microbial Growth Dynamics Coupled to Nutrient and Oxygen Transport in Porous Media," *Water Resour. Res.*, 22:107.

41 Widdowson, M. A., F. J. Molz and L. D. Benefield. 1987. "Development and Application of a Model for Simulating Microbial Growth Dynamics Coupled to Nutrient and Oxygen Transport in Porous Media," in *Proc. AGWSE/IGWMCH Conference on Solving Ground Water Problems with Models, Denver, CO*, National Water Well Association, Dublin, OH, pp. 28–51.

42 Widdowson, M. A., F. J. Molz and L. D. Benefield. 1988. "A Numerical Transport Model for Oxygen- and Nitrate-Based Respiration Linked to Substrate and Nutrient Availability in Porous Media (Paper 7W5057)," *Water Resources Research*, 24(9):1553–1565.

43 Srinivasan, P. and J. W. Mercer. 1988. "Simulation of Biodegradation and Sorption Processes in Ground Water," *Ground Water*, 26(4):475–487.

44 MacQuarrie, K. T. B., E. A. Sudicky and E. O. Frind. 1990. "Simulation of Biodegradable Organic Contaminants in Groundwater, (1), Numerical Formulation in Principal Direction," *Water Resour. Res.*, 26(2):207–222.

45 MacQuarrie, K. T. B. and E. A. Sudicky. 1990. "Simulation of Biodegradable Or-

ganic Contaminants in Groundwater, (2), Plume Behavior in Uniform and Random Flow Fields," *Water Resour. Res.*, 26(2):223–239.

46 Rifai, H. S. and P. B. Bedient. 1987a. "BIOPLUME II—Two Dimensional Modeling for Hydrocarbon Biodegradation and In Situ Restoration," in *Proc., NWWA Conference on Petroleum Hydrocarbons and Organic Chemicals in Ground Water, Houston, TX,* pp. 431–450.

47 Rifai, H. S., P. B. Bedient, R. C. Borden and J. F. Haasbeek. 1987b. "BIOPLUME II. Computer Model of Two-Dimensional Contaminant Transport Under the Influence of Oxygen Limited Biodegradation in Ground Water," *User's Manual Version 1.0,* National Center for Ground Water Research, Rice University, Houston, TX.

48 Konikow, L. F. and J. D. Bredehoeft. 1978. "Computer Model of Two-Dimensional Solute Transport and Dispersion in Ground Water, Automated Data Processing and Computations," *Techniques of Water Resources Investigations of the U.S. Geological Survey,* Washington, D.C.

CHAPTER 7

Soil Vapor Extraction Systems

VACUUM extraction is an important new technology that has emerged to treat spills of volatile organic compounds in the unsaturated zone. It is suited to removing volatile nonaqueous phase liquids (NAPLs) in the unsaturated zone that are trapped at residual saturation, and free product layers. Experience has shown that in comparison to remedial strategies like excavation or pump-and-treat, soil venting can be more efficient and cost-effective [11,12]. Vacuum extraction involves passing large volumes of air through or close to a contaminated spill using an air circulation system. The organic compounds or various fractions of a mixture of organic compounds volatilize or evaporate into the air and are transported to the surface. Thus, just like the other contaminant transport problems considered earlier in the book, the vapor extraction method involves fluid flow (air in this case) and the transport of mass dissolved in the fluid.

Many complex processes occur on the microscale. However, the three main factors that control the performance of a venting operation are the chemical composition of the contaminant, vapor flow rates through the unsaturated zone, and the flow path of carrier vapors relative to the location of the contaminants.

Among the advantages of the soil air extraction process are that it creates a minimal disturbance of the contaminated soil, it can be constructed from standard equipment, there is demonstrated experience with the process at pilot- and field-scale, it can be used to treat larger volumes of soil than are practical for excavation, and there is a potential for product recovery. With vapor extraction, it is possible to clean up spills before the chemicals reach the groundwater table. Soil vapor extraction technology is often used in conjunction with other cleanup technologies to provide complete restoration of contaminated sites [7,14,16].

Unfortunately, there are few guidelines for the optimal design, installation, and operation of soil vapor extraction systems [4]. Theoretically based design equations that define the limits of this technology are especially lacking. Because of this, the design of these systems is mostly empirical. Alternative designs can only be compared by the actual construction, operation, and monitoring of each design. A large number of pilot- and full-scale soil vapor extraction systems have been constructed and studied under a wide range of conditions. The information gathered from this experience has been used to deduce the effectiveness of this technology in a major state of the technology review [10].

SYSTEM COMPONENTS

A typical soil vapor extraction system such as the one shown in Figure 7.1 consists of: (1) one or more extraction wells, (2) one or more air inlet or injection wells (optional), (3) piping or air headers, (4) vacuum pumps or air blowers, (5) flow meters and controllers, (6) vacuum gauges, (7)

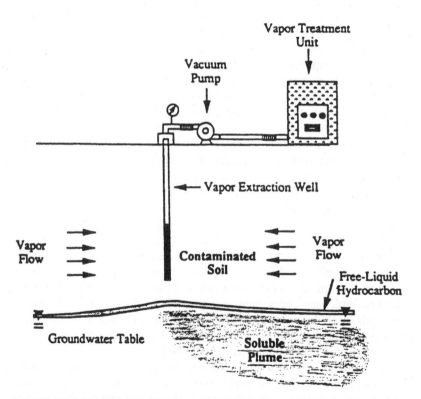

Figure 7.1 "Basic" in situ soil venting system (reprinted by permission of Ground Water Monitoring Review, copyright 1990, worldwide rights reserved).

sampling ports, (8) air/water separator (optional), (9) vapor treatment (optional), and (10) a cap (optional).

Extraction wells are typically designed to fully penetrate the unsaturated zone to the capillary fringe. If the groundwater is at a shallow depth or if the contamination is confined to near-surface soils, then the extraction wells may be placed horizontally. Extraction wells usually consist of slotted, plastic pipe placed in permeable packing. The surface of the augured column for vertical wells or the trench for horizontal wells is usually grouted to prevent the direct inflow of air from the surface along the well casing or through the trench.

It may be desirable to also install air inlet or injection wells to control air flow through zones of maximum contamination. They are constructed similarly to the extraction wells. Inlet wells or vents are passive and allow air to be drawn into the ground at specific locations. Injection wells force air into the ground and can be used in closed-loop systems [17]. The function of inlet and injection wells is to enhance air movement in strategic locations and promote horizontal air flow to the extraction wells. Pipes and headers may be buried or wrapped with heat tape in insulated northern climates to prevent freezing of condensate.

The pumps or blowers reduce gas pressure in the extraction wells and induce air flow to the wells. The pressure from the outlet side of the pumps or blowers can be used to push the exit gas through a treatment system and back into the ground if injection wells are used. Gas flow meters are installed to measure the volume of extracted air. Pressure losses in the overall system are measured with vacuum gauges. Sampling ports may be installed in the system at each well head, at the blower, and after vapor treatment. In addition, vapor and pressure monitoring probes may be placed to measure soil vapor concentrations and the radius of influence of the vacuum in the extraction wells.

To protect the blowers or pumps and to increase the efficiency of vapor treatment systems, an air/water separator may need to be installed. The condensate may then have to be treated as a hazardous waste depending on the types and concentrations of contaminants. The need for a separator may be eliminated by covering the treatment area with an impermeable cap or by designing the extraction wells to separate water from air within the well packing. An impermeable cap serves to cover the treatment site to minimize infiltration and controls the horizontal movement of inlet air.

Vapor treatment may not be required if the emission rates of chemicals are low or if they are easily degraded in the atmosphere. Typical treatment systems include liquid/vapor condensation, incineration, catalytic conversion, or granular activated carbon adsorption.

Patterns of air circulation to extraction wells have been studied in the field by direct measurements [2], and more recently by mathematical and experimental modeling [11,12,13]. Most of the theoretical work to date has assumed that any density differences in the vapor can be neglected under the forced convective conditions created by the vacuum extraction. The following section presents the governing equations that describe the flow and transport of vapor in the subsurface.

GOVERNING EQUATIONS OF FLOW

Johnson et al. [11,12] have exploited the analogy to groundwater flow in developing simple screening models to describe the distribution in pressure around venting wells. For conditions of radial flow, the governing equation can be written

$$\frac{1}{r}\frac{\partial}{\partial r}\left(r\frac{\partial P'}{\partial r}\right) = \left(\frac{\epsilon\mu}{kP_{atm}}\right)\frac{\partial P'}{\partial t} \tag{7.1}$$

where P' is the deviation of pressure form the reference pressure P_{atm}, k is soil permeability, μ is the vapor viscosity, ϵ is porosity, and t is time. When Equation (7.1) is solved with appropriate boundary conditions, with m as the thickness of the unconfined zone and r as the radial distance from the well to the point of interest,

$$P' = \frac{Q}{4\pi m(k/\mu)} W(u) \qquad \text{where } u = r^2\epsilon\mu/4kP_{atm}t \tag{7.2}$$

where $W(u)$ is the well function of u, which is a commonly tabulated function.

Calculations with Equation (7.2) show that for sandy soils ($10 < k < 100$ darcys) the pressure distribution approximates a steady state within hours. Thus, it is appropriate to model pressure distributions using a steady-state solution to the governing flow equation. For the following set of boundary conditions: $P = P_w$ at $r = R_w$ and $P = P_{atm}$ at $r = R_I$, where P_w is the pressure at the well with radius R_w and P_{atm} is the ambient pressure at the radius of influence R_I.

Johnson et al. [11] provide the following solution to the steady-state equation for radial flow

$$P(r) = P_w\left\{1 + \left[1 - \left(\frac{P_{atm}}{P_w}\right)^2\right]\frac{ln(r/R_w)}{ln(R_w/R_I)}\right\} \tag{7.3}$$

As Johnson et al. [11] point out, while not explicitly represented in Equation (7.3), the properties of the soil do influence the steady-state pressure distribution because the radius of influence (R_I) does vary as a function of permeability and layering. Johnson et al. [11] develop corresponding solutions for radial darcy velocity and volumetric flow rate for this steady-state case. The latter of these solutions provides a useful way to determine what the theoretical maximum air flow is to a vapor extraction well and is written

$$\frac{Q}{H} = \pi \frac{k}{\mu} P_w \frac{[1 - (P_{atm}/P_w)^2]}{1n(R_w/R_I)} \tag{7.4}$$

where H is the total length of the screen. Just as is the case with well hydraulics, these kinds of analytical equations not only form the basis for predictive analysis, but also for well testing methods for permeability estimations.

Analytical solutions are most useful for screening purposes and for exploring the relationships among variables; however, their practical applicability is limited to simple problems. An alternative way of solving differential equations like Equation (7.1) is with powerful numerical models like the one developed by Krishnayya et al. [13]. Numerical models are effective in modeling complicated problems of the type commonly encountered in practice.

By analogy with problems involving groundwater flow to wells, the general pattern of air circulation will be influenced by features of the fluid being circulated, the geologic system, and the well system used to withdraw and inject air. The permeability is the most important of all the parameters that influence flow. Ultimately, it is permeability that determines the efficacy of vapor extraction because flow rates at steady state for a well under a specified vacuum are a direct function of permeability. Vapor extraction, to be practically useful, requires some minimal rate of air circulation, which may not be feasible in some low permeability units.

From an analytical model, Johnson et al. [12] developed a series of relationships between permeability and flow rate (Figure 7.2). For a given vacuum in the extraction well, the steady-state rate of air flow is a linear function of permeability (log scales). An increase in the vacuum (smaller P_w) at a given permeability will increase the air flow. However, the maximum change that might be expected is about an order of magnitude in flow rate. For estimation purposes, a value of $R_I = 12$ m (40 ft) can be used without a significant loss of accuracy. Figure 7.2 presents predicted flow rates per unit well screen thickness Q/H, expressed in "standard" volumetric units $Q*/H$ ($= Q/H(P_w/P_{atm})$ for a 4-in diameter extraction well, and a wide range of soil permeabilities and applied vacuums.

It is well known from groundwater ·flow theory that variability in per-

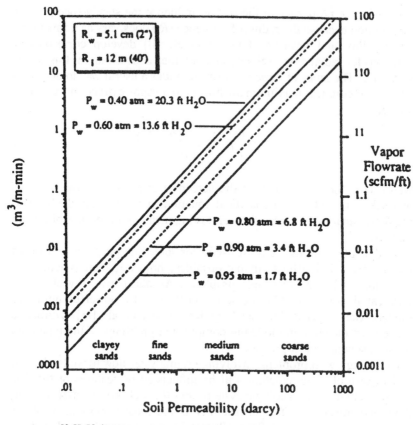

[ft H₂O] denote vacuums expressed as equivalent water column heights

Figure 7.2 Predicted steady-state flow rates (per unit well screen thickness) for a range of soil permeabilities and applied vacuums (P_w) (reprinted by permission of Ground Water Monitoring Review, copyright 1990, worldwide rights reserved).

meability plays an important role in controlling the pattern of flow. This is also the case for patterns of air circulation. Most circulation will occur through the most permeable zones for a layered system, and this can have a profound effect on the rate and efficiency of cleanup schemes.

Features of the design of the system as a whole also control air circulation. The most important factors in this respect include: (1) the flow rates in the injection/extraction wells, (2) the types and locations of wells in a multiwell system, and (3) the presence of a surface seal. For a single vacuum well, Figure 7.2 illustrates how reducing P_w increases the air withdrawal rate. Similarly, adding more wells to the system will also do the same. Furthermore, the pattern of air circulation can be controlled by the types and locations of the wells.

ORGANIC VAPOR TRANSPORT

The vapor phase transport of mass has much in common with the transport of dissolved contaminants in groundwater. The processes of advection and dispersion operate to physically transport the mass, while the chemical processes are involved with the generation of the contaminants through volatilization and subsequent interactions with the water and solid phases in the system.

Removal of mass from the spill depends upon the process of volatilization, which is a phase partitioning between a liquid and a gas. For solvents, the process is described in terms of equilibrium theory by a form of Raoult's law

$$P_i = x_i t_i P_i^0 \tag{7.5}$$

where P_i is the vapor pressure of component i (atm) in the soil gas, x_i is the mole fraction of the component in the solvent, and P_i^0 is the vapor pressure of the pure solvent at the temperature of interest; these are tabulated constants [20]. An activity coefficient t_i for the ith component in the mixture is added to account for nonidealities [9]. By applying the ideal gas equation, the partial pressures calculated in Equation (7.5) can be expressed in terms of concentration or

$$C_s = \sum_i \frac{x_i P_i MW_i}{RT} \tag{7.6}$$

where C_s is the saturation concentration in mg/L, T is temperature in degrees Kelvin, R is the universal gas constant (0.0821 1-atm/mole^{-k}), MW_i is the number of milligrams per mole, and P_i is the partial pressure of the compound.

The equilibrium model provides the basis for most analyses carried out in practice. The equilibrium approach applies in cases where the rate of volatilization is large relative to the rate of physical transport through the medium. Where the flow of air contacts residual product in every pore, phase equilibrium is achieved quite rapidly at least for the more volatile compounds at the start of venting. One important reason for less than equilibrium concentrations of vapor is that geologic heterogeneities cause most of the air flow to avoid most of the contaminated zones.

With spills of complex solvent mixtures like gasoline, soil venting removes the components with higher vapor pressure first. The residual contamination thus becomes progressively enriched in the less volatile compounds. Because of this compositional change, the overall rate of mass removal decreases with time. This process has been described mathematically by Marley and Hoag [15] and Johnson et al. [11] using forms of Equation (7.6) and information on the most volatile components of gasoline.

Vapor pressures for organic compounds increase significantly as a function of increasing temperature. For example, the vapor pressure for a volatile compound like benzene increases from 0.037 atm at 32°F to 0.137 atm at 80°F [11]. Over the same temperature range, the vapor pressure for n-dodecane increases 2.8×10^{-5} atm at 2.3×10^{-4} atm. The main implication of this result is that the overall time required for cleanup will change depending upon temperature.

The physical transport process, advection and dispersion, play an important role with respect to vapor transport. The simplest and most extensively analyzed model of physical transport assumes that advection transports contaminants from the point of generation to the vapor extraction well [Figure 7.3(a)]. The distribution of contaminants would be reasonably homogeneous and much of the air flow would have the opportunity to move through the bulk of the spill. Ideally, the vapor concentration in that fraction of the air moving through the spill is the equilibrium concentration determined by Raoult's law. In Figure 7.3(a), about 75% of the air passes through the spill, giving a vapor concentration in the well that is 75% of equilibrium value and a removal rate of $0.75QC_{tot}$.

However, in heterogeneous media or when pure product is present, the air flow may not pass directly through the spill. Specific examples cited by Johnson et al. [11,12] include: (1) air flow across the surface of a free liquid floating on the water table or low permeability layer [Figure 7.3(b)], or (2) product trapped in a lower permeability lens [Figure 7.3(c)]. In both of these cases, the mass loss rate of contaminants is controlled by the rate at which mass can diffuse into the moving vapor stream. Thus, when flowing air bypasses the spill, the rate of mass removal may be much lower than for the homogeneous case. The result may be a vapor extraction system that circulates considerable quantities of air without removing much of the contaminant. Mathematical approaches for estimating the contaminant removal rates under these more complex conditions are presented by Johnson et al. [12].

ILLUSTRATIVE EXAMPLE OF VAPOR EXTRACTION COMPUTATIONS

Johnson et al. [11] examined a hypothetical case in which 400 gallons (approximately 1500 liters) of gasoline were spilled into 1412 ft³ of soil at a 10% moisture content. This spill provides a residual saturation of 2% gasoline by dry weight or 20,000 ppm total petroleum content. Other conditions specified include: a soil bulk density of 1.5 gm/cm³, a porosity of 0.40, an organic carbon fraction (f_{oc}) of 0.01, an air flow rate of 20 ft³/min,

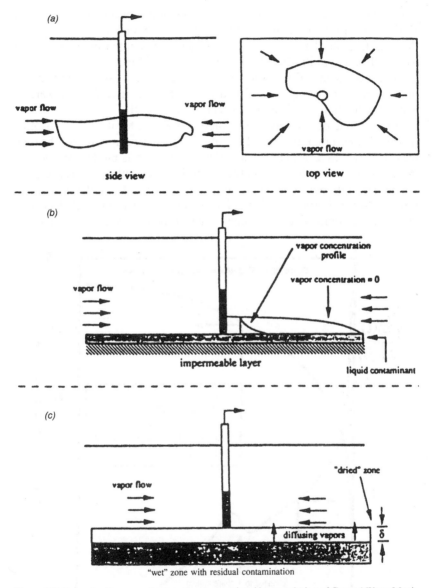

(a)

vapor flow   vapor flow

vapor flow

side view   top view

(b)

vapor concentration
profile

vapor concentration = 0

vapor flow

impermeable layer

liquid contaminant

(c)

"dried" zone

vapor flow

δ

diffusing vapors

"wet" zone with residual contamination

Figure 7.3 Scenarios for removal rate estimates (reprinted by permission of Ground Water Monitoring Review, copyright 1990, worldwide rights reserved).

a soil temperature of 60°F, and a relative humidity for the incoming air of 100%. The problem assumes that 25% of the circulating air passes through the spill. Thus, component concentrations in the vapor extraction well will be at equilibrium for an air flow of 5 ft³/m, which is the value used in subsequent calculations. Also required for the calculation is information about the compounds found in gasoline, their molecular weights and mass fractions, and their physical properties [11].

The total rate of mass loss from vapor extraction for the entire gasoline mixture is the sum of the mass loss rates $C_i Q$ for the compounds. The total mass loss rate as a ratio of the initial mass loss rate is plotted versus time in Figure 7.4. Also depicted in the figure is the cumulative percentage of the initial spill recovered as a function of time. As contaminants are removed during venting, the residual soil contamination level decreases and mixture compositions become richer in the less volatile compounds.

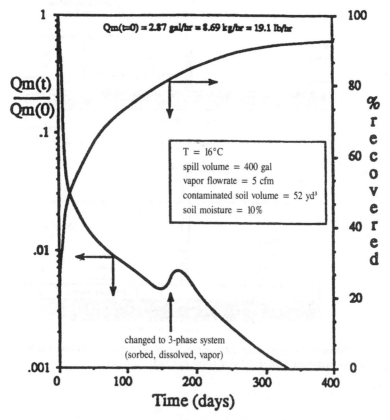

Figure 7.4 Predicted mass loss rates for a hypothetical venting operation (reprinted by permission of NWWA, copyright 1988, worldwide rights reserved).

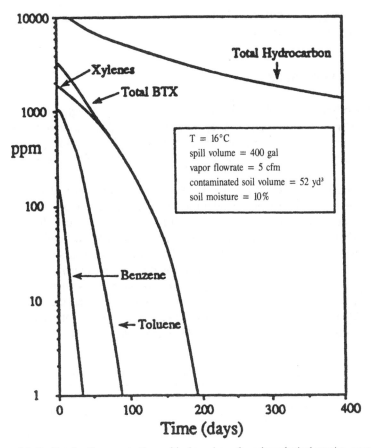

Figure 7.5 Predicted soil concentrations of hydrocarbons for a hypothetical venting operation (concentrations represent mass of hydrocarbons/mass of dry soil) (reprinted by permission of NWWA, copyright 1988, worldwide rights reserved).

Both of these processes result in decreased vapor concentrations, and hence, decreased removal rates with time.

The example shows clearly how the rate of mass loss decreases with time as the most volatile components are removed from the mixture. Figure 7.5 displays the time variation in the soil concentrations of a number of the most volatile compounds in gasoline. The pattern of removal rates (i.e., benzene > toluene > xylenes) is explained by the respective vapor pressures (i.e., 0.10 > 0.029 > 0.0066 to 0.0088 atm). At the end of 400 days, the residual product consists mainly of the larger molecular weight compounds in gasoline, having vapor pressures in general less than 0.005 atm. Within 200 days the BTEX level is reduced to below 1 ppm, while the total hydrocarbon concentration is reduced to about 1000 ppm. Johnson et al. [12] present a detailed design approach for a service station remediation.

SYSTEM VARIABLES

A number of variables characterize the successful design and operation of a vapor extraction system. They may be classified as site conditions, soil properties, chemical characteristics, control variables, and response variables [1,8]. Table 7.1 lists specific variables that belong to these groups [10].

The extent to which VOCs are dispersed in the soil, vertically and horizontally, is an important consideration in deciding if vapor extraction is preferable to other methods. If the spill has penetrated more than twenty or thirty feet or has spread through an area over several hundred square feet at a particular depth, or if the spill volume is in excess of 500 cubic yards, then excavation costs begin to exceed those associated with a vapor extraction system [6,17]. The depth to groundwater is also important. Where groundwater is at depths of more than forty feet and the contamination extends to the groundwater, use of soil vapor extraction systems may be one of the few ways to remove VOCs from the soil [14]. Groundwater

TABLE 7.1. **Soil Vapor Extraction System Variables.**

Site Conditions	*Control Variables*
Distribution of VOCs	Air withdrawal rate
Depth to groundwater	Well configuration
Infiltration rate	Extraction well spacing
Location of heterogeneities	Vent well spacing
Temperature	Ground surface covering
Atmospheric pressure	Pumping duration
	Inlet air VOC concentration
Soil Properties	and moisture content
Permeability (air and water)	
Porosity	*Response Variables*
Organic carbon content	Pressure gradients
Soil structure	Final distribution of VOCs
Soil moisture characteristics	Final moisture content
Particle size distribution	Extracted air concentration
	Extracted air moisture
Chemical Properties	Extracted air temperature
Henry's constant	Power usage
Solubility	
Adsorption equilibrium	
Diffusivity (air and water)	
Density	
Viscosity	

depth in some cases may be lowered to increase the volume of the unsaturated zone.

Heterogeneities influence air movement as well as the location of chemicals, and the presence of heterogeneities make it more difficult to position extraction and inlet wells. There generally will be significant differences in the air conductivity of the various strata of a stratified soil. A horizontally stratified soil may be favorable for vapor extraction because the relatively impervious strata will limit the rate of vertical inflow from the ground surface and will tend to extend the influence of the applied vacuum horizontally from the point of extraction.

The soil characteristics at a particular site will have a significant effect on the applicability of vapor extraction systems. Air conductivity controls the rate at which air can be drawn from soil by the applied vacuum. Grain size, moisture content, soil aggregation, and stratification probably are the most important properties [3,10]. The soil moisture content or degree of saturation is also important in that it is easier to draw air through drier soils. As the size of a soil aggregate increases, the time required for diffusion of the chemical out of the immobile regions also increases. However, even clayey or silty soils may be effectively ventilated by the usual levels of vacuum developed in a soil vapor extraction system [5,18]. The success of the soil vapor extraction in these soils may depend on the presence of more conductive strata, as would be expected in alluvial settings, or on relatively low moisture content in the finer-grained soils.

Chemical properties will dictate whether a soil vapor extraction system is feasible. A vapor-phase vacuum extraction system is most effective at removing compounds that exhibit significant volatility at the ambient temperatures in soil. Low molecular weight volatile compounds are favored, and vapor extraction is likely to be most effective at new sites where the more volatile compounds are still present. It has been suggested that compounds exhibiting vapor pressures over 0.5 mm of mercury can most likely be extracted with soil air [3]. When expressed in terms of the air/water partitioning coefficient, compounds that have values of dimensionless Henry's Law constants greater than 0.01 are more likely to be removed in vapor extraction systems. Examples of compounds that have been effectively removed by vapor extraction include trichloroethene, trichloroethane, tetrachloroethene, and most gasoline constituents. Compounds that are less applicable to removal include trichlorobenzene, acetone, and heavier petroleum fuels [3,17,19].

Soil vapor extraction processes are flexible in that several variables can be adjusted during design or operation. These variables include the air withdrawal rate, the well spacing and configuration, the control of water infiltration by capping, and the pumping duration. Higher air flow rates tend to increase vapor removal because the zone of influence is increased and air is forced through more of the air-filled pores. More wells will allow better control of air flow but will also increase construction and operation costs. Intermittent operation of the blowers will allow time for chemicals to diffuse from immobile water and air and permit removal at higher concentrations.

Parameters responding to soil vapor extraction system performance include: air pressure gradients, VOC concentrations, moisture content, and power usage. The rate of vapor removal is expected to be primarily affected by the chemical's volatility, its sorptive capacity onto soil, the air flow rate, the distribution of air flow, the initial distribution of chemicals, soil stratification or aggregation, and the soil moisture content.

DESIGN ISSUES ON VAPOR EXTRACTION SYSTEMS

Johnson et al. [12] present a detailed and comprehensive approach that can be followed in practice for implementing a vapor extraction scheme. Having arrived at vapor extraction as a candidate strategy for remediating a contamination problem, the following four major activities would constitute a vapor extraction project:

(1) Feasibility analysis based on available data to establish whether a vapor system is an appropriate remedial strategy

(2) Testing with a pilot vapor extraction system and groundwater wells to determine physical parameters and confirm the accuracy of above analysis

(3) Design of the complete system

(4) Monitoring to confirm that the designed level of cleanup is being achieved

The feasibility analysis requires preliminary information about the geology and hydrogeology of the site, the chemical and physical characteristics of the liquid contaminant, and any regulations applicable to the spill. As Table 7.2 indicates, there are in general five questions that need to be answered. Most of the approaches to answering the questions have been discussed in this section to some extent.

In terms of ongoing monitoring activities, Johnson et al. [12] recommend that the following data be collected on an ongoing basis: (1) air flow rates, (2) pressure at each extraction and injection well, (3) temperature, ambient and soil, (4) water table elevation, (5) vapor concentrations and composition at each extraction well, and (6) soil-gas vapor concentrations at various distances from the extraction wells. The vapor extraction system will be shut down when target levels of a cleanup are reached. These

TABLE 7.2. **A Summary of the Basic Issues to Be Addressed in Planning for the Development of a Vapor Extraction System.**

Major Activity/Issue	Approach to Resolving Issue
Feasibility analysis: (1) What is the expected range in rates of air flow for a single well?	Estimation of Q with equations.
(2) What is the estimated removal rate with a single well?	Estimate total connection of vapor in the air flow C_{tot}. Estimate single well removal rate (percent air flow contacting spill $\times Q \times C$).
(3) Is the single well removal rate acceptable?	Compare removal rate for a single well with the rate required (i.e., total mass of spill $\times$ target time for recovering spill).
(4) What residual contamination will remain at the end and will the level of cleanup meet regulatory requirements?	Given the air flow rates and a model describing the removal rates as a function of time, determine soil concentrations of key constituents at the end of the remedial period. Compare these levels with requirements of the regulations.
(5) What negative effects may develop as a consequence of vapor extraction?	Evaluate the possibility of off-site contributions from other sources. Evaluate the possibility of a water table rise accompanying vapor extraction.
Testing with vapor extraction and groundwater wells:	Evaluation of pumping tests gives site-specific data necessary for finalizing the design of the vapor extraction and water control systems. Monitoring of vapor concentrations during the well test helps to validate earlier calculations and establish whether efficiencies are affected by heterogeneities.
Final system design:	Site-specific data is used in conjunction with equations to determine the single well removal rate. The required removal rate (1 c this table) is divided by the single well rate to establish the number of wells.
Monitoring:	Monitoring documents the progress of the cleanup and will assist in determining whether changes to the system are required. Monitoring helps to establish when the system can be shut off.

targets are often site-specific, depending upon particular water quality and health standards and safety considerations. Johnson et al. [12] discuss in detail how the monitoring data are evaluated in making this decision.

CONCLUSIONS

Based on the current state of the technology of soil vapor extraction systems [10] and the extensive review from Johnson et al. [11,12], the following conclusions can be made regarding soil vapor systems:

(1) Soil vapor extraction can be effectively used for removing a wide range of volatile chemicals over a wide range of conditions.

(2) The design and operation of these systems is flexible enough to allow for rapid changes in operation, thus improving the removal of chemicals.

(3) Intermittent blower operation may be more efficient in terms of removing the most chemical with the least energy, especially in systems where chemical transport is limited by diffusion through air or water. However, this may not be the most cost-effective means of remediation because total cleanup times may be significantly larger.

(4) Volatile chemicals can be extracted from clays and silts but at a slower rate.

(5) Air injection and capping a site have the advantage of controlling air movement, but injection systems need to be carefully designed.

(6) Extraction wells are usually screened from a depth of from five to ten feet below the surface to the groundwater table. For thick zones of unsaturation, maximum screen lengths of twenty to thirty feet are specified.

(7) Air/water separators are simple to construct and should probably be installed in most systems.

(8) Installation of a cap over the area to be vented reduces the chance of extracting water and extends the path that air follows from the ground surface, thereby increasing the volume of soil treated.

(9) Incremental installation of wells, while probably more expensive, allows for a greater degree of freedom in design. Modular construction, where the most contaminated zones are vented first, is preferable.

(10) Use of soil vapor probes in conjunction with soil borings to assess final cleanup is less expensive than use of soil borings alone. It is usually impossible to do a complete materials balance on a given site because most sites have an unknown amount of VOC on the soil and in the groundwater.

(11) Soil vapor extraction systems are usually only part of a much larger site remediation system.

There are still many other technical issues that need to be resolved in the future. The usefulness of forced or passive vapor injection wells is often debated, as well as other means of controlling vapor flow paths (impermeable surface covers, for example). A well-documented demonstration of the effectiveness of soil venting for the removal of contaminants from low-permeability soils is also needed. It is clear from the simplistic modeling results presented that venting will be less effective in such situations. Without a comparison with other viable treatment alternatives, however, it is difficult to determine if soil venting would still be the preferred option in such cases.

REFERENCES

1 Anastos, G. J., P. J. Marks, M. H. Corbin and M. F. Coia. 1985. "Task 11. In Situ Air Stripping of Soils, Pilot Study, Final Report," Report No. AMXTH-TE-TR-85026, U.S. Army Toxic and Hazardous Material Agency, Aberdeen Proving Grounds, Edgewood, MD, p. 88.

2 Batchelder, G. V., W. A. Panzeri and H. T. Phillips. 1986. "Soil Ventilation for the Removal of Adsorbed Liquid Hydrocarbons in the Subsurface," *Petroleum Hydrocarbons and Organic Chemicals in Ground Water—Prevention, Detection, and Restoration*, National Water Well Association. Dublin, OH, pp. 672–688.

3 Bennedsen, M. B., J. P. Scott and J. D. Hartley. 1985. "Use of Vapor Extraction Systems for In Situ Removal of Volatile Organic Compounds from Soil," *Proceedings of National Conference on Hazardous Wastes and Hazardous Materials*, HMCRI, pp. 92–95.

4 Bennedsen, M. B. 1987. "Vacuum VOCs from Soil," *Pollution Engineering*, 19(2):66–68.

5 Camp, Dresser, and McKee, Inc. 1987. "Interim Report for Field Evaluation of Terra Vac Corrective Action Technology at a Florida LUST Site," Contract No. 68-03-3409, U.S. Environmental Protection Agency, Edison, NJ.

6 CH₂M-Hill, Inc. 1985. "Remedial Planning/Field Investigation Team, Verona Well Field—Thomas Solvent Company, Battle Creek, Michigan," Operable Unit Feasibility Study, Contract No. 68-01-6692, U.S. Environmental Protection Agency, Chicago, IL.

7 CH₂M-Hill, Inc. 1987. "Operable Unit Remedial Action, Soil Vapor Extraction at Thomas Solvents Raymond Road Facility, Battle Creek, MI," Quality Assurance Project Plan, U.S. Environmental Protection Agency, Chicago, IL.

8 Enviresponse, Inc. 1987. "Demonstration Test Plan, In-Situ Vacuum Extraction Technology, Terra Vac Inc., SITE Program, Groveland Wells Superfund Site, Groveland, MA," EERU Contract No. 68-03-3255, Work Assignment 1-R18, Enviresponse No. 3-70-06340098, Edison, NJ.

9 Hinchee, R. E., H. J. Reisinger, D. Burris, B. Marks and J. Stepek. 1986. "Underground Fuel Contamination, Investigation, and Remediation—a Risk Assessment Approach to How Clean is Clean," *Petroleum Hydrocarbons and*

Organic Chemicals in Ground Water—Prevention, Detection, and Restoration, National Water Well Association, Dublin, OH, pp. 539–563.

10 Hutzler, N. J., B. E. Murphy and J. S. Gierke. 1989. "State of Technology Review Soil Vapor Extraction Systems," Cooperative Agreement No. CR-814319-01-1, Risk Reduction Engineering Lab, U.S. Environmental Protection Agency, Cincinnati, OH.

11 Johnson, P. C., M. W. Kemblowski and J. D. Colthart. 1988. "Practical Screening Models for Soil Venting Applications," *Petroleum Hydrocarbons and Organic Chemicals in Ground Water—Prevention, Detection, and Restoration*, National Water Well Association, Dublin, OH, pp. 521–546.

12 Johnson, P. C., C. C. Stanley, M. W. Kemblowski, D. L. Byers and J. D. Colthart. 1990. "A Practical Approach to the Design, Operation, and Monitoring of In-Situ Soil Venting Systems," *Ground Water Monitoring Review*, 10:2:159–178.

13 Krishnayya, A. V., M. J. O'Connor, J. G. Agar and R. D. King. 1988. "Vapor Extraction Systems—Factors Affecting their Design and Performance," *Petroleum Hydrocarbons and Organic Chemicals in Ground Water—Prevention, Detection and Restoration*, National Water Well Association, Dublin, OH, pp. 547–569.

14 Malot, J. J. and P. R. Wood. 1985. "Low Cost, Site Specific, Total Approach to Decontamination," *Conference on Environmental and Public Health Effects of Soils Contaminated with Petroleum Products*, University of Massachusetts, Amherst, MA.

15 Marley, M. C. and G. E. Hoag. 1984. "Induced Soil Venting for the Recovery/Restoration of Gasoline Hydrocarbons in the Vadose Zone," *Petroleum Hydrocarbons and Organic Chemicals in Ground Water—Prevention, Detection and Restoration*, National Water Well Association, Dublin, OH, pp. 473–503.

16 Oster, C. C. and N. C. Wenck. 1988. "Vacuum Extraction of Volatile Organics from Soils." *Proceedings of the 1988 Joint CSCE-ASCE National Conference on Environmental Engineering*, Vancouver, B.C., Canada, pp. 809–817.

17 Payne, F. C., C. P. Cubbage, G. L. Kilmer and L. H. Fish. 1986. "In Situ Removal of Purgeable Organic Compounds from Vadose Zone Soils," *Purdue Industrial Waste Conference*.

18 Terra Vac, Inc. 1987. "Union 76 Gas Station Clean-up, Bellview, Florida," Florida Department of Environmental Regulation, Tallahassee, FL.

19 Texas Research Institute. 1980. "Examination of Venting for Removal of Gasoline Vapors from Contaminated Soil," American Petroleum Institute (Reprinted 1986).

20 Verschueren, K. 1983. *Handbook of Environmental Data on Organic Chemicals*. Van Nostrand Reinhold, New York, NY.

CHAPTER 8

Multiphase Contamination and Free Product Recovery

NONAQUEOUS phase liquids (NAPLs) are known to be present at numerous industrial and waste disposal sites, and they are suspected to exist at many more. Due to the numerous variables influencing NAPL transport and fate in the subsurface, they are likely to go undetected and yet they are likely to be a significant limiting factor in site remediation. This is especially true for Dense NAPLs (DNAPLs), such as chlorinated solvents, for which unfortunately there are few proven remediation technologies available [1]. This chapter focuses on the variables that govern the subsurface transport and fate of NAPLs, and in particular on monitoring and recovery of Light NAPLs (LNAPLs) during groundwater remediation.

PRINCIPLES OF MULTIPHASE FLOW

A nonaqueous phase liquid (NAPL) is a term used to describe the physical and chemical differences between a hydrocarbon liquid and water that result in a distinct physical interface that will separate the two phases in a mixture. The interface, which is visually observable, divides the bulk phases of the two liquids, but individual compounds may solubilize from the NAPL into the groundwater. NAPLs are generally divided into two classes based on their densities. Those that are less dense than water (LNAPLs) will pool and spread as a floating free product layer upon the water table if they are released to the subsurface in sufficient quantities. Typical examples of LNAPLs are gasoline and most petroleum hydrocarbon mixtures. Those nonaqueous phase liquids with densities greater than water (DNAPLs) can pass across the water table and may be found at great depths within the saturated zone of a groundwater aquifer. DNAPLs are typically chlorinated solvents such as trichloroethene (TCE) and tetrachloroethene (PERC), wastes from wood preserving processes (creosote, pentachlorophenol), coal tars, and pesticides.

The features that make transport and fate characteristics of NAPLs unique from those of miscible constituents are directly or indirectly associated with the presence of the interface separating the two phases. These features are also present in the unsaturated zone where the immiscible fluids are air and water. The characteristics of multiphase flow have been studied for some time by petroleum engineers and by soil scientists. Only recently have environmental engineers been drawn to this subject as they address problems related to subsurface contamination and remediation. The important principles of multiphase flow are presented first from the standpoint of the air/water system in the unsaturated zone. More general multiphase systems will be discussed later.

CAPILLARITY

For an unsaturated medium, part of the pore space is filled with water while the rest is filled with air. The sum of the volumetric water and air contents is equal to the total porosity:

$$\theta_w + \theta_a = n \tag{8.1}$$

where θ_w is the volumetric water content, θ_a is the volumetric air content, and n is the porosity of the soil. The new processes that occur in unsaturated media are associated with the presence of the interface separating one fluid phase from another. Consider an interface separating the water and air phases as shown in Figure 8.1. For a molecule at point A within the interior of the water phase, the cohesive forces between molecules pull equally in all directions and there is no net force. For a molecule at point B near the surface, the cohesive forces result in a net force toward the interior of the water phase. To move a molecule from A to B we must break the neighbor bonds and move against this force field. The result is that molecules at the interface have more energy than molecules within the bulk phase. The excess surface energy, σ, is called the interfacial energy (ergs/cm²), or surface tension (dynes/cm).

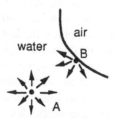

Figure 8.1 Cohesive forces between molecules within the interior of a fluid and at the interface.

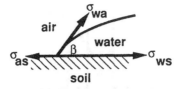

Figure 8.2 Contact angle used to classify wetting characteristics.

When a solid phase is present the contact angle, β, as shown in Figure 8.2, is of great interest because it determines the wetting properties of the porous matrix. In Figure 8.2, there are three phases present along with three interfaces, and each has its own surface tension. The surface tensions and contact angle are related through Young's equation (see Adamson [2]), which is one of the fundamental equations in the theory of capillarity:

$$\cos \beta = \frac{\sigma_{as} - \sigma_{ws}}{\sigma_{wa}} \tag{8.2}$$

where σ_{ij} is the surface tension between phases i and j. The main feature of Young's equation is that it shows that the contact angle is a thermodynamic function of the interfacial energies of the phases present, and thus the contact angle itself is a thermodynamic property that may be used to distinguish the wetting characteristics of the fluids present with respect to the solid phase. If the contact angle between the solid and the interface for a phase is less than 90° then the phase is said to be the wetting phase. The phase with contact angle greater than 90° is the nonwetting phase. For most soils, water is the wetting fluid, even with respect to a hydrocarbon nonwetting phase.

The water rise in a capillary tube of radius r is shown in Figure 8.3. The capillary forces pulling the column upward must be balanced by the weight of the water column in the tube. This leads to the relation

$$\Psi = \frac{2\sigma_{wa} \cos \beta}{\gamma r} \tag{8.3}$$

where Ψ is the capillary rise.

Equation (8.3) shows that the amount of capillary rise is directly proportional to the interfacial tension and inversely proportional to the radius of the capillary tube. Water in naturally occurring earth materials most often has a value of $\sigma_{wa} \cos \beta$ of about 60 dynes/cm, whereas σ_{wa} for pure water in contact with air at 20°C is about 72 dynes/cm.

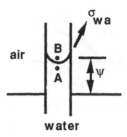

Figure 8.3 Rise of the air/water interface in a capillary tube caused by surface tension.

The pressure difference across the interface separating the wetting and nonwetting phases is called the capillary pressure, and it is defined by

$$p_c = p_{nw} - p_w \qquad (8.4)$$

For the capillary shown in Figure 8.3, this capillary pressure is given by

$$p_c = \frac{2\sigma_{wa} \cos \beta}{r} \qquad (8.5)$$

Equation (8.5) is an example of a much more general result. Any time there is an interface that is curved, then there is an associated pressure difference across that interface. This pressure difference is given by

$$p_c = \frac{2\sigma}{r_c} \qquad (8.6)$$

where r_c is the average radii of curvature. Equation (8.6) is called Laplace's equation, and it is also fundamental to the theory of capillarity.

Leverett [3] has investigated how the capillary pressure varies with elevation under conditions of vertical static equilibrium. Choosing an elevation datum where $p_c = 0$, he finds

$$p_c = \Delta \varrho g z \qquad (8.7)$$

where $\Delta \varrho = \varrho_w - \varrho_{nw}$ is the density difference between the wetting and nonwetting phases, Equation (8.7) shows that the capillary pressure increases with distance above the datum. For the air/water system the density of air is negligible compared to that of water. Thus $\Delta \varrho \cong \varrho_w$ and with the same datum chosen in Equation (8.7) one has the capillary pressure head given by

$$\frac{p_c}{\varrho_w g} = \Psi = z \qquad (8.8)$$

Equation (8.8) shows that the capillary pressure head, Ψ, is equal to the elevation above a datum where p_c equals zero. Equation (8.6), in turn, suggests that the datum is the same as that of the capillary rise with a large-diameter capillary, such as that found in an observation well. Since there is essentially no capillary rise in a typical observation well, the datum corresponds to the elevation of the water table.

The capillary pressure increases with elevation above the water table. Laplace's Equation (8.6) shows that the mean radius of curvature of the interfaces separating the wetting and nonwetting phases decreases with increasing capillary pressure, and thus with elevation above the water table. This suggests that the interfaces are moving into smaller and smaller pores. Thus the wetting phase saturations are decreasing while the nonwetting phase saturations are increasing with elevation above a free fluid interface. These arguments suggest that at higher capillary pressures, the water/air interfaces are pulled out of the larger pores and remain present only in the smaller pores. Thus, at higher elevations above the water table, water is retained in the smaller pores while the larger pores contain air. A schematic view of the pore water distribution is shown in Figure 8.4, where the bounding curve is the pore size "density" function (fraction of pores at a given size).

SOIL WATER CHARACTERISTIC CURVE

The soil water characteristic curve provides the relationship between the capillary pressure and water content for a particular soil. The distribution shown in Figure 8.4 of water and air within the pores corresponds to a particular capillary pressure. The integral of the water-filled pore fraction corresponds to the fraction of the total porosity which is filled with water. This is the volumetric water content. Thus the relationship between p_c and θ is a function of the pore size distribution. The size of the pores containing air/water interfaces at a given capillary pressure is given roughly by Equation (8.6).

The soil water characteristic for most soils shows hysteresis. This means that the p_c-θ relationship depends on the saturation history as well as the

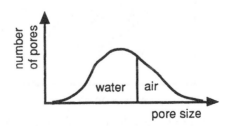

Figure 8.4 Phase content as a function of pore size.

existing water content. A typical characteristic curve is shown in Figure 8.5. The upper curve corresponds to a soil sample that starts off initially saturated and is drained by increasing the capillary pressure. At first a relatively small number of the pores are drained with increasing p_c because there are a limited number of large pores. There is a rapid decrease in water content with increasing p_c through the range of most plentiful pore size. At very large p_c there is little change in water content with increasing suction. Ultimately, increasing the suction cannot remove more water. The saturation at which this occurs is called the irreducible water saturation. The lower curve corresponds to a rewetting of the soil. A different sequence of pores is filled, corresponding in each case to the next largest pore, rather than the next smallest pore as is the case during drainage. Ultimately, insular saturation is reached and the air phase becomes trapped in the largest pores and the wetting process ceases. From this point onward there is no unique relationship between the water pressure and the water content. If the wetting stops or drainage process stops anywhere between the endpoint limits and is reversed, then the scanning curves would be followed. What actually happens in the field during a rainfall event is that the soil initially becomes wetted, following a wetting scanning curve, and then turns around to start to drain, following a drainage scanning curve. If wetting or drainage occurs for a long enough time period, then the primary wetting or drainage curves are reached.

Soil water characteristic curves can be measured in the laboratory using either (1) displacement methods based on the establishment of successive states of hydrostatic equilibrium or (2) dynamic methods based on the es-

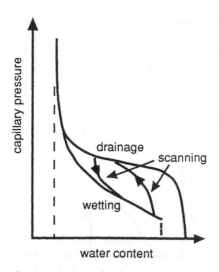

Figure 8.5 Soil water characteristic curve showing drainage, wetting, and scanning curves.

tablishment of successive states of steady flow of a pair composed of a wetting and a nonwetting fluid. For a discussion of these methods see Bear [4], Corey [5], and Dullien [6].

For many applications it is convenient to fit parametric models of the soil water characteristic to laboratory or field data. A major advantage of these parameter models is that there has been a large amount of field data collected and fit to these models. These data provide an estimate of the parameters and their expected range based solely on the soil texture. This is particularly helpful because the type of information that is often available deals with the soil texture, and one can often make a good estimate of the parameters without having to resort to more complicated laboratory testing methods. There are a number of parametric models that have been suggested in the literature. The two most widely used models at present are the power-law model of Brooks and Corey [7] and the model suggested by van Genuchten [8]. Both of these models are for fitting the drainage curve.

The Brooks and Corey [7] model takes the form of a power-law relating the effective or reduced saturation, Θ, to the capillary pressure head, Ψ:

$$\Theta = \begin{cases} 1 & \Psi \leq \Psi_b \\ \left(\dfrac{\Psi_b}{\Psi}\right)^\lambda & \Psi > \Psi_b \end{cases} \tag{8.9}$$

The reduced saturation is defined by

$$\Theta = \frac{\theta - \theta_r}{n - \theta_r} \tag{8.10}$$

In Equation (8.10), θ is the volumetric water content and θ_r is the irreducible water content. The capillary pressure head is defined by Equation (8.8). The parameters appearing in the Brooks and Corey model are the bubbling capillary pressure head (also called the displacement pressure head), Ψ_b, and the pore size distribution index, λ.

The van Genuchten [8] model takes the form

$$\Theta = \left(\frac{1}{1 + (\alpha\Psi)^N}\right)^M \tag{8.11}$$

for $\Psi \geq 0$. The parameters in the van Genuchten model are α, N, and M. With Mualem's [9] relative permeability model, the parameters N and M are related through $M = 1 - 1/N$ or $N = 1/(1 - M)$. For large capillary heads, the Brooks and Corey and the van Genuchten models become identical if $\lambda = MN$ and $\Psi_b = 1/\alpha$.

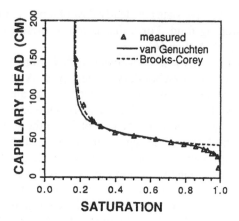

Figure 8.6 Comparison of fitted Brooks and Corey and van Genuchten models with data for a fine sand.

Figure 8.6 compares the Brooks and Corey and the van Genuchten models with measured data for a fine sand. The irreducible saturation $(s_r = \theta_r/n)$ is estimated to be $s_r = 0.167$. The Brooks and Corey model parameters are $\Psi_b = 41$ cm and $\lambda = 3.74$, while the parameters for the van Genuchten model are $\alpha = 0.0205$ cm^{-1}, $M = 0.853$, and $N = 6.80$. The fit of both models is very good. As expected, the Brooks and Corey model is less accurate near $s \approx 1$, where the model itself has a discontinuous slope. At larger Ψ, the Brooks and Corey model has a somewhat better fit, though throughout the range, the van Genuchten model has an overall better fit to the measured data. From this single test it is not possible to judge which is the better model. The van Genuchten model has the definite advantage of being able to fit the data throughout the entire saturation range. On the other hand, the van Genuchten model leads to rather difficult equations for analysis of saturation and flow relations. The Brooks and Corey model is very simple analytically, as are its permeability representations (discussed below). While it is less accurate at high saturations, it may be the preferred model for some applications where high saturation conditions are unlikely to occur.

DARCY'S LAW FOR UNSATURATED FLOW

In steady multiphase flow, the fluids flow practically independently of each other as the interfacial boundaries between them are mostly situated in capillaries where there is no flow (Dullien [6]). Hence, neither fluid influences the flow behavior of the other one, and each fluid flows in the capillary network allotted to it, just as if it were the only fluid present. The analogous form of Darcy's law for the unsaturated flow of water is

$$\mathbf{U} = -\frac{k_e}{\mu} (\nabla p + \varrho g \mathbf{k}) \tag{8.12}$$

which is the same as Equation (2.6) in Chapter 2 except that in place of the intrinsic permeability k, the "effective permeability" k_e appears, which is a function of the water saturation or volumetric content. The extension of Darcy's law given by Equation (8.12) appears to have been first suggested by Muskat and co-workers (Muskat and Meres [10], Wyckoff and Botset [11], Muskat et al. [12]). Because of capillary pressure hysteresis, the effective permeabilities, in general, depend on the saturation history. It is customary to express the effective permeabilities as fractions of the intrinsic permeability, k, of the medium. This defines the relative permeability as

$$k_r = \frac{k_e}{k} \tag{8.13}$$

and the extension of Darcy's law may be written

$$\mathbf{U} = -K(\theta) \operatorname{grad}(h) \tag{8.14}$$

or

$$\mathbf{U} = -K_s k_r(\theta) \nabla h \tag{8.15}$$

where h is the usual hydraulic head given by

$$h = z - \Psi \tag{8.16}$$

and K_s is the saturated hydraulic conductivity.

The relative permeability of water varies from 1 under saturated conditions to 0 at the irreducible water content. The $k_r(\theta)$ relation does not appear to show much hysteresis. With the soil water characteristic, one can also write the relative permeability as a function of the capillary pressure head. However, the $k_r(\Psi)$ function does show considerable hysteresis.

The relative permeability function can be measured in the laboratory, though these measurements are difficult. An alternative approach is to use the pore size distribution to estimate the relative permeability function of the media. Such an approach is conceptually based on Figure 8.4. This figure shows that the smaller pores are filled with the wetting fluid (water) while the larger pores are filled with the nonwetting fluid (air). If an effective permeability can be assigned to pores of a given size, through Poiseuille's law for example, then one may integrate across the range of pores which are filled with a given fluid to assign a medium permeability to that fluid. The Poiseuille formula shows that the fluid velocity in a capillary tube is proportional to its radius squared, while the Laplace equation shows that the pore radius is inversely proportional to the capillary head. Concepts and models such as these allow one to derive theories for prediction of the relative permeability function from a measured or estimated capillary pressure curve. Perhaps the most widely used models at present are those of Burdine [13] and Mualem [9].

Brooks and Corey [7] used their power law characteristic model of Equation (8.9) in the Burdine equations to derive the wetting (water) and nonwetting (air) relative permeabilities as

$$k_{rw} = \Theta^{(3+2/\lambda)} \tag{8.17}$$

and

$$k_{ra} = (1 - \Theta)^2(1 - \Theta^{(1+2/\lambda)}) \tag{8.18}$$

Equation (8.17) is often written as

$$k_{rw} = \Theta^\epsilon \tag{8.19}$$

where $\epsilon = e + 2/\lambda$.

Brooks and Corey [7] also present experimental data reporting the wetting and nonwetting phase relative permeability. These data, along with the theoretical curves of Equations (8.18) and (8.19) with parameters estimated from the fit of the characteristic curve from Figure 8.6, are shown in Figure 8.7. A major point of interest is that the permeability data was not fit to the models. Rather, the model parameters that are derived from fitting the soil water characteristic data are used in the theoretical permeability models. This implies that one may be able to get by using the laboratory measurement of the soil water characteristic curve in lieu of the much more difficult relative permeability measurements in the laboratory.

MONITORING OF FLOATING FREE PRODUCT HYDROCARBONS

Groundwater contamination due to surface spills or subsurface leakage of petroleum hydrocarbons and other nonaqueous phase liquids (NAPLs)

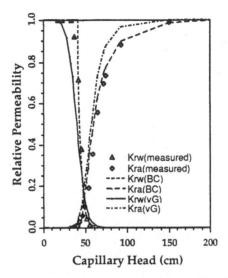

Figure 8.7 Comparison of the Brooks and Corey model and the van Genuchten model for relative permeability with parameters estimated from the capillary pressure curves.

is a widespread problem. In a recent review of EPA groundwater extraction remedies, 112 sites were surveyed (EPA [14]). The results showed that 95% of the sites were contaminated by organics and at 14% of the sites, the remediation objectives included nonaqueous liquid recovery. In addition, permeable geologic materials were present at most of the sites with over 80% of the surveys noting the presence of sand and gravel.

While EPA Superfund sites often show contamination with solvents and other DNAPLs, which tend to continue sinking when reaching the water table, at sites associated with the petroleum industry (e.g., refineries and service stations), one is more likely to find NAPLs with densities less than that of water. Upon reaching the water table, these LNAPLs will float and tend to spread laterally along the capillary fringe. At some locations, vast quantities of free product hydrocarbons are found floating at the water table, and control and remediation of these sites is of major environmental interest.

Mathematical models have been developed which have the potential to significantly improve the design and analysis of free floating product systems for control and recovery of hydrocarbons. These models may be used to determine the best locations for recovery wells and the optimal water and hydrocarbon production rates. The models may also be used to estimate the unrecoverable hydrocarbon volumes (both free product and residual) as well as the potential offsite migration of the floating product. This section concerns results that are of interest in monitoring floating free product and estimating free product volumes from monitoring data. The following section will consider the optimal oil and water production rates for recovery systems.

VERTICAL DISTRIBUTION OF FLOATING FREE PRODUCT HYDROCARBONS

When a LNAPL reaches the water table as free product it depresses the water table under its own weight and eventually spreads laterally due to energy gradients within its phase. The distribution of free product is determined by its density relative to that of water, by the total amount of hydrocarbon present, and by capillary pressure forces acting within the porous matrix. Under static conditions the free product will reach a condition of vertical equilibrium where the pressure gradient within both the water and hydrocarbon phase satisfy the hydrostatic pressure equation. Under these conditions the actual distribution of free product may be determined from the pore size distribution, as summarized by the water retention curve of the soil using the method first outlined by Schiegg [15]. This method directly relates the free product distribution within the porous matrix to the product thickness one would observe in a well in free communication with the aquifer fluids. This is shown schematically in Figure 8.8. Within the formation the fluids form a three-phase system. However, in the observation well one finds three distinct fluid layers—a water layer below, an LNAPL layer floating on the water, and air above. Since the observation well thickness is independent of capillary forces and

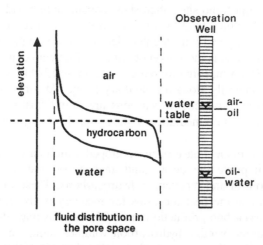

Figure 8.8 Distribution of LNAPLs within a porous formation and the apparent thickness seen in an observation well.

corresponds to the same energy distribution as the formation fluids under conditions of vertical equilibrium, this observation well thickness provides a useful analog for estimation of free product volumes present within the aquifer and for computation of free product recovery by pumped wells. Such a modeling approach assumes that the fluids are always in vertical equilibrium and calculates the lateral energy gradients from the changes in elevation and free product thickness that would be seen in an observation well that could at any time be brought into equilibrium with the formation fluids. It should be noted, however, that formations fluids need not actually exist under conditions of vertical equilibrium. Indeed, Kemblowski and Chiang [16] have shown that under dynamic conditions the observation well thickness of a free product need show little relation to the formation distribution of fluids, but rather is a strong function of whether the water table is rising or falling.

The theory for using the observation well thickness to estimate the actual quantity of hydrocarbon liquid present in the formation is based on the following concepts. First, the air/water soil characteristic curve provides sufficient information describing the pore size distribution for the soil. Second, the characteristic curve for a different fluid system may be estimated from the air/water curve using scaling parameters that depend on properties of the fluids involved. Third, in a three-phase system, the capillary pressure between the NAPL and water phases is a function of the water (wetting fluid) content, while the total liquid content (NAPL plus water) is a function of the interfacial curvature at the air/oil interface. This third assumption was suggested by Leverett [4] and has been supported by experiment. With these assumptions one can predict the vertical distribution of fluids in a three-phase system, and from this distribution, estimate the amount of hydrocarbon present for a given observation well thickness. The major results from this theory as they apply to the power-law parametric model are outlined below.

One may apply the theory of capillarity in multiphase systems to estimate free product volumes by noting that the soil water characteristic curve, as shown in Figure 8.5, gives the vertical distribution of water in a two-phase (air/water) system. In terms of the power-law parametric model equations, this is provided explicitly by combining Equation (8.8) and (8.9) to find

$$\Theta = \begin{cases} 1 & z \leq \Psi_b \\ \left(\dfrac{\Psi_b}{z}\right)^{\lambda} & z > \Psi_b \end{cases} \tag{8.20}$$

where z is measured upward from the water table and the reduced saturation, Θ, is given by Equation (8.10). Equation (8.20) gives the volumetric water content as a function of elevation above the water table under conditions of vertical equilibrium. According to this power-law model, the pore size distribution index along with the bubbling pressure (a measure of the largest pore size) serve to determine the structure of the porous media. In order to apply this model for a multiphase system including a free product hydrocarbon, one must determine how the equilibrium behavior for an air/oil or oil/water system can be estimated from those for a air/water system, where "oil" (Oily Immiscible Liquid) may be considered an alternative acronym for the floating free product hydrocarbon (LNAPL). If changes in soil structure (swelling, etc.) are neglected, the difference in behavior from one fluid system to another can be attributed only to differences in fluid properties. Assume that the characteristic parameters for the air/water system are given: Ψ_b, λ, n, and θ_r. For multiple fluid systems use the subscripts "w," "o," and "a" to designate the water, oil, and air phases. In addition, for fluid pairs, the subscript order is nonwetting fluid first and wetting fluid second. Thus, for the air/water system Equation (8.20) is written

$$\Theta_w = \left(\frac{\Psi_{baw}}{z} \right)^{\lambda} \tag{8.21}$$

The question of interest concerns how to generalize this result for different fluid systems. In general, for the ij-system, the bubbling pressure is related to that for the aw-system through

$$p_{bij} = p_{baw} \frac{\sigma_{ij}}{\sigma_{aw}} = \varrho_w g \Psi_{baw} \frac{\sigma_{ij}}{\sigma_{aw}} \tag{8.22}$$

which follows because the maximum pore size remains constant and the bubbling pressure depends only on the surface tension. Since $p_{cij} = \Delta\varrho_{ij}gz$ (where again, the datum is chosen at the elevation where the capillary pressure vanishes), this gives

$$\Theta_j = \left(\frac{\varrho_w \Psi_{baw} \sigma_{ij}}{\Delta\varrho_{ij}\sigma_{aw}z} \right)^{\lambda} = \left(\frac{\Psi_{bij}}{z} \right)^{\lambda} \tag{8.23}$$

for $z \geq \Psi_{bij}$ where j is the wetting phase and

$$\Psi_{bij} = \frac{\varrho_w \Psi_{baw} \sigma_{ij}}{\Delta\varrho_{ij}\sigma_{aw}} \tag{8.24}$$

Similar scaling relationships were introduced by Leverett [4] and later used by van Dam [17], Schiegg [15], Parker et al. [18], Cary et al. [19], Demond and Roberts [20], and others. For the air/oil system one may assume $\Delta \varrho_{ao} = \varrho_o$ since the density of air is small.

In a three-phase system, water is considered to be the wetting fluid, oil is of intermediate wettability, and air is the nonwetting fluid. The implication of this is that water will reside in the smallest pores, air in the largest pores, and oil in the intermediate pore sizes if all three fluids are present within a sample. Since capillary pressure relations are only for two-fluid systems, one has to work with the fluid pairs separately in a three-phase system. This approach has been developed by Leverett [4] and adopted by Schiegg [15] and by Parker et al. [18]. The basic assumption is that the total liquid content, $\theta_t = \theta_w + \theta_o$, in a multiphase system is a function of the interfacial curvature at the air/oil interface, independent of the number or proportions of liquids contained in the porous medium. It is also assumed that the water saturation in a three-phase system depends only on the oil/water capillary pressure. With the power-law retention model this may be written

$$\Theta_w = \Theta_w(p_{cow}) = \left(\frac{\Psi_{bow}}{z - z_{ow}} \right)^{\lambda} \tag{8.25}$$

$$\Theta_t = \Theta_t(p_{cao}) = \left(\frac{\Psi_{bao}}{z - z_{ao}} \right)^{\lambda} \tag{8.26}$$

where z_{ow} and z_{ao} are the elevations at which the corresponding capillary pressures would vanish.

If one wishes to consider the case where the oil may reside at possibly differing residual saturations above and below the free product region near the water table, the appropriate scaling functions for the reduced saturations are

$$\Theta_w(p_{cow}) = \frac{\theta_w - \theta_{wr}}{n - \theta_{wr} - \theta_{ors}} \tag{8.27}$$

$$\Theta_t(p_{cao}) = \frac{\theta_t - \theta_{wr} - \theta_{orv}}{n - \theta_{wr} - \theta_{orv}} = \frac{\theta_w + \theta_o - \theta_{wr} - \theta_{orv}}{n - \theta_{wr} - \theta_{orv}} \tag{8.28}$$

where θ_{wr} is the water retention or "field capacity," and θ_{ors} and θ_{orv} are the residual oil contents in the saturated and vadose zones, respectively.

Together, Equations (8.25) through (8.28) determine the fluid distribution near the water table. What is still lacking is a determination of the capillary pressure datums z_{ow} and z_{ao}. However, these are the levels at which one would find the fluid interfaces in observation wells where capillary forces are absent, and the problem reduces to the standard manometer problem from hydrostatics. Let the elevation z_{aw} be that of the free water interface in the absence of oil, while z_{ao} and z_{ow} are the corresponding elevations when an oil layer of apparent thickness b_o and density ϱ_o is present. A simple calculation from hydrostatics shows that

$$z_{aw} - z_{ow} = \frac{\varrho_o}{\varrho_w} b_o \qquad (8.29)$$

where ϱ_w is the density of water. One also finds that

$$z_{ao} - z_{aw} = \left(\frac{\varrho_w - \varrho_o}{\varrho_w} \right) b_o \qquad (8.30)$$

For example, if $b_o = 5$ ft and $\varrho_o = 0.8$ gm/cc, then from Equations (8.29) and (8.30) one finds that $z_{ao} - z_{aw} = 1$ ft while $z_{aw} - z_{ow} = 4$ ft. In addition, since the fluid head is given by $h = p/\gamma + z$, one has

$$h_o = z_{ao} \qquad (8.31)$$

or $h_o = 1$ ft, and $h_w = 0$ ft (where it is assumed that the air pressure is atmospheric). More generally, for conditions of vertical equilibrium, the head in the water phase is given by

$$h_w = \frac{p_{ow}}{\varrho_w g} + z_{ow} = \left(1 - \frac{\varrho_o}{\varrho_w} \right) z_{ow} + \frac{\varrho_o}{\varrho_w} z_{ao} \qquad (8.32)$$

where p_{ow} is the fluid pressure at the oil/water interface. It should be noted that z_{ao} and z_{ow} are the fluid levels that would be found in an observation well in equilibrium with the fluids in the formation with an oil head h_o and water head $h_w = 0$. The thickness of oil observed in the well, b_o, is given by

$$b_o = z_{ao} - z_{ow} = \left(\frac{\varrho_w}{\varrho_w - \varrho_o} \right) h_o \qquad (8.33)$$

which comes from Equations (8.29) and (8.30). The relations from Equations (8.29) through (8.33) also hold true for observation well thicknesses of free product because here too, capillary forces are not important and conditions of vertical equilibrium hold. Similar relationships are presented by van Dam [17].

Equations (8.25) through (8.30) provide the fluid distributions within the free product zone given an apparent observation well thickness b_o. Fig-

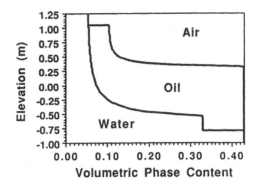

Figure 8.9 Gasoline distribution in a sand soil corresponding to a 1 m depth in an observation well.

ure 8.9 shows an example for gasoline ($\varrho_o = 0.718$ gm/cc) with an observation well thickness of $b_o = 1$ m. The assumed soil is a sand with $n = 0.43$, $\theta_{wr} = 0.045 \cong 0.05$, $\Psi_{baw} = 0.069$ m, and $\lambda = 1.68$. These represent the mean values from 246 samples of a sand soil reported by Carsel and Parrish [21]. It is also assumed that $\theta_{orv} = 0.05$ and $\theta_{ors} = 0.10$. (Typical residual saturations vary from 0.10 to 0.20 in the vadose zone and from 0.15 to 0.50 in the saturated zone, Mercer and Cohen [22]). With the further assumptions that $\sigma_{aw} = 65$ dynes/cm, $\sigma_{ow} = 45$ dynes/cm, and $\sigma_{ao} = 35$ dynes/cm, then Equation (8.24) gives $\Psi_{bow} = 0.169$ m and $\Psi_{bao} = 0.052$ m. With the 1 m oil layer thickness and the reported density of gasoline, Equations (8.29) and (8.30) give $z_{ow} = -0.718$ m and $z_{ao} = 0.282$ m. Also shown in Figure 8.9 are residual saturations of thickness 0.25 m above and below the free product thickness. These correspond to a situation where the water table fluctuates over a range of 0.5 m during a cycle (annually or tidal cycle) with the current conditions corresponding to mid-cycle.

The total thickness of hydrocarbon present in the free product region (not including the hydrocarbon possibly present at residual saturations above or below the free product) is found by integrating the difference between the total liquid content and the water content over the free product region:

$$D_o = \int (\theta_t - \theta_w)dz \tag{8.34}$$

This usage of the oil layer thickness, D_o, corresponds to that of Schwille [23], who uses it for the ratio between the amount of oil spreading laterally on the groundwater surface and the area occupied by it. Other authors refer to the oil layer thickness as that which may be visually observed in a laboratory apparatus. With Equations (8.25) through (8.28) one may evaluate the integral given by Equation (8.34). The result may be written

$$D_o = \alpha + \beta(b_o)b_o \qquad (8.35)$$

where

$$\alpha = \frac{[\lambda(n - \theta_{wr}) - \theta_{ors}]\Psi_{bow} - [\lambda(n - \theta_{wr} - \theta_{orv})]\Psi_{bao}}{1 + \lambda} \qquad (8.36)$$

and

$$\beta(b_o) = (n - \theta_{wr}) + \frac{\chi}{1 - \chi}\theta_{orv} - \frac{(n - \theta_{wr} - \theta_{ors})}{(1 - \lambda)}\left(\frac{(1 - \chi)\Psi_{bow}}{b_o}\right)^{\lambda} \qquad (8.37)$$

and where

$$\chi = \frac{\sigma_{ao}}{\sigma_{ow}}\left(\frac{\varrho_w - \varrho_o}{\varrho_o}\right)\left(\frac{n - \theta_{wr} - \theta_{orv}}{n - \theta_{wr} - \theta_{ors}}\right)^{1/\lambda} \qquad (8.38)$$

A similar result has been presented by Parker and Lenhard [24]. For the distribution shown in Figure 8.9, Equation (8.35) gives $D_o = 0.324$ m. In Equation (8.35), $\beta(b_o)$ has only a weak dependence on b_o, especially at moderate to large oil layer thicknesses. This implies that the relationship between D_o and b_o is nearly linear.

A representative curve for the same fluid system is shown in Figure 8.10. This figure is for a soil with $n = 0.4$, $\theta_{wr} = 0.04$, $\Psi_{baw} = 0.23$ ft, $\varrho_o = 0.72$ gm/cc, $\theta_{orv} = 0.05$, $\theta_{ors} = 0.10$, $\sigma_{ao} = 35$ dynes/cm, and $\sigma_{ow} = 45$ dynes/cm. Similar curves may be developed for other soil and fluid systems, though these results are representative. Roughly, the actual formation thickness varies from one-quarter to one-third the observation well thickness. In general one finds that the slope of the D_o versus b_o curve decreases for decreasing λ (finer textured soils with a wider range of pore sizes), but Figure 8.10 suggests that that dependency is slight.

RECOVERY OF FREE PRODUCT THROUGH DUAL PUMP SYSTEMS

The discussion now turns to recovery of hydrocarbons through use of pumping systems. There are a number of types of pumping systems available for recovery of floating free product hydrocarbon from an aquifer. When there is only a small amount of free product present, then the

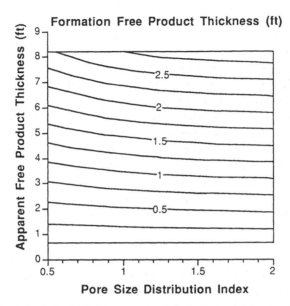

Figure 8.10 Formation LNAPL thickness as a function of soil texture (pore size distribution index) and observation well LNAPL thickness.

simplest of these is to place a pump in a well that will skim off any NAPL that enters the well. These are typically low production rate systems, and when significant quantities of hydrocarbon are present, they are not very efficient. An alternative dual pump system is shown in Figure 8.11. These dual pump systems are most useful when there may be a significant amount of LNAPL present. The pump with its intake below the hydrocar-

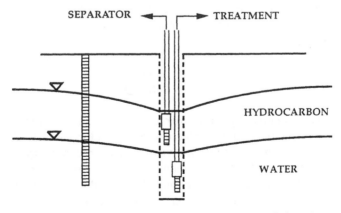

Figure 8.11 Dual pump system for recovery of free product hydrocarbon.

bon layer produces water with dissolved hydrocarbon constituents. Its effluent generally will require treatment before ultimate disposal or reinjection. Production of water creates a "bowl" or depression in the water table into which the hydrocarbon can collect for removal. In addition, water production decreases the tendency of the oil/water interface to upcone, influencing the hydrocarbon removal efficiency. The pump with its intake within the hydrocarbon layer will produce mostly free product, though water may also be produced, and the effluent will then be sent to an API separator before treatment of the aqueous phase. Alternately, the hydrocarbon well may be a skimmer well that will produce essentially only NAPL that enters the well.

The fact that there are two pumping units present in a dual well system, both of which may produce liquid at independent rates, means that there is an optimal production rate for NAPL which can be achieved from a well. The water production rate determines the overall extent of lowering of the water table near the well. Use of a large water production rate will allow a corresponding large NAPL recovery rate because the large depression in the water table formed by the water production will cause a large gradient in the NAPL layer and a larger flow of NAPL to the well. However, a significant lowering of the water table will smear the hydrocarbon over a larger volume of the aquifer, leaving behind a larger residual of NAPL that cannot be removed by conventional recovery methods.

For a given water production rate, there is also an optimal hydrocarbon recovery rate that may be achieved. Obviously, if the NAPL production rate is set at a low level, then little hydrocarbon will be recovered. On the other extreme, if the NAPL well production rate is set too high, then upconing of the water level in the well will occur and the well will pump off and produce little hydrocarbon. The optimal rate will be an intermediate rate where the hydrocarbon flows to the well at a rate that directly matches the gradient caused by the water production well. Numerical simulation studies have suggested that the optimal hydrocarbon production rate may be estimated from

$$Q_o = \frac{2\pi T_o s}{\ln (R/r_w)} \tag{8.39}$$

where Q_o is the optimal hydrocarbon recovery rate, s is the drawdown that would be induced by a water production rate Q_w in the absence of the free product, T_o is the effective transmissivity of the hydrocarbon layer floating on top of the water table, R is the radius of influence of the water production well, and r_w is the effective radius of the production well. Equation (8.39) suggests that the hydrocarbon recovery rate is proportional to the water production rate (through the magnitude of s) and it is also proportional to the saturated water hydraulic conductivity, soil texture, and saturated hydrocarbon layer thickness through the parameter T_o. A reasonable estimate of $\ln(R/r_w)$ is $\ln(R/r_w) \cong 7$.

To place Equation (8.39) in a more useful form one may note that

$$T_o = K_w \frac{\mu_w}{\mu_o} \frac{\varrho_o}{\varrho_w} b_o k_{ro} \tag{8.40}$$

where K_w is the hydraulic conductivity of the aquifer to water, μ_i and ϱ_i are the respective phase dynamic viscosities and densities, and μ_w/μ_o, ϱ_o/ϱ_w is the ratio of the kinematic viscosity of water to oil, b_o is the oil layer thickness and k_{ro} is the effective oil layer relative permeability. The kinematic viscosity ratio varies from about 0.5 for 30 degree API oil to 0.1 for 40 degree API oil. The effective relative permeability varies with the hydrocarbon distribution in the aquifer and can be estimated from the soil texture and the apparent oil layer thickness. To estimate the drawdown, s, one may use the Dupuit equation, which is

$$H^2 - h_w^2 = \frac{Q_w}{\pi K_w} \ln\left(\frac{R}{r_w}\right) \tag{8.41}$$

and the drawdown is then given as $s = H - h_w$. This equation may be approximated as

$$s \text{ (ft)} = \frac{Q_w \text{ (gpm)}}{500 \, K_w \text{ (cm/sec)}} \tag{8.42}$$

The resulting optimal hydrocarbon production rate for a unit drawdown of the water table and a unit kinematic viscosity ratio is shown in Figure 8.12. This figure was developed from Figure 8.10 with the relative permeability estimated from the Burdine [13] equation.

As an example of use of Figure 8.12, consider a recovery well with a 5 ft layer of 35 API NAPL in a 40 ft thick fine sand aquifer with $K_w = 0.005$ cm/s and a water production rate of 15 gpm. According to Figure 8.12, the optimal production rate is 0.35 gpm. This value has to be corrected for the actual drawdown and NAPL kinematic viscosity. From Equation (8.42), the drawdown caused by a water production rate of 15 gpm is about 6 ft. For 35 API hydrocarbon the kinematic viscosity ratio is about 0.22. Thus the optimal hydrocarbon production rate is $0.35 \times 6 \times 0.22 = 0.46$ gpm. Thus an NAPL production rate of about 0.5 gpm can be achieved, giving a ratio of water to oil production of 30.

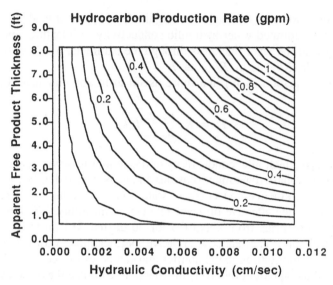

Figure 8.12 Optimal hydrocarbon production rate for a unit drawdown of the water table (ft) and for a unit kinematic viscosity ratio.

REFERENCES

1 Huling, S. G. and J. W. Weaver. 1991. "Dense Nonaqueous Phase Liquids," *Ground Water Issue*, United States Environmental Protection Agency, Office of Research and Development, EPA/540/4-91/002.

2 Adamson, A. W. 1978. *Physical Chemistry of Surfaces, 3rd Ed.* Interscience, New York.

3 Leverett, M. C. 1941. "Capillary Behavior in Porous Media," *Trans. AIME*, 142:341–358.

4 Bear, J. 1972. *Dynamics of Fluids in Porous Media.* Elsevier, New York (1988, Dover edition available).

5 Corey, A. T. 1977. *Mechanics of Heterogeneous Fluids in Porous Media.* Water Resources Publications, Fort Collins, CO.

6 Dullien, F. A. L. 1979. *Porous Media: Fluid Transport and Pore Structure.* Academic Press, New York.

7 Brooks, R. H. and A. T. Corey. 1964. "Hydraulic Properties of Porous Media," *Hydrol. Pap. 3*, Colo. State Univ., Fort Collins, CO.

8 van Genuchten, M. T. 1980. "A Closed-Form Equation for Predicting the Hydraulic Conductivity of Unsaturated Soil," *Soil Sci. Soc. Am. J.*, 44:892–898.

9 Mualem, Y. 1974. "A New Model for Predicting the Hydraulic Conductivity of Unsaturated Porous Media," *Water Resour. Res.*, 12:513–522.

10 Muskat, M. and M. W. Meres. 1936. *Physics*, 7:346.

11 Wyckoff, R. D. and H. G. Botset. 1936. "The Flow of Gas-Liquid Mixture through Unconsolidated Sands," *Physics*, 7:325–345.

12 Muskat, M., Wyckoff, R. D., Botset, H. G. and M. W. Meres. 1937. "Flow of Gas Liquid Mixtures through Sands," *Trans. A.I.M.E. Petrol.*, 123:69–96.

13 Burdine, N. T. 1953. "Relative Permeability Calculations from Pore-Size Data," *Trans. A.I.M.E.*, 198:71–77.

14 United States Environmental Protection Agency. 1989. "Evaluation of Ground-Water Extraction Remedies, Volume 1, Summary Report," Office of Emergency and Remedial Response, Washington, D.C., September, EPA/540/2-89/054.

15 Schiegg, H. O. 1985. "Consideration on Water, Oil and Air in Porous Media," *Water Science and Technology*, 17:467–476.

16 Kemblowski, M. W. and C. Y. Chiang. 1990. "Hydrocarbon Thickness Fluctuations in Monitoring Wells," *Ground Water*, 28:244–252.

17 van Dam, J. 1967. "The Migration of Hydrocarbons in a Water-Bearing Stratum," in *The Joint Problems of the Oil and Water Industries*, P. Hepple, ed., The Institute of Petroleum, London.

18 Parker, J. C., R. J. Lenhard and T. Kuppusamy. 1987. "A Parametric Model for Constitutive Properties Governing Multiphase Flow in Porous Media," *Water Resour. Res.*, 23:618–624.

19 Cary, J. W., C. S. Simmons and J. F. McBride. 1989. "Permeability of Air and Immiscible Organic Liquids in Porous Media," *Water Resources Bulletin*, 25(6):1205–1216.

20 Demond, A. H. and P. V. Roberts. 1991. "Effect of Interfacial Forces on Two-Phase Capillary Pressure-Saturation Relations," *Water Resour. Res.*, 27(3): 423–437.

21 Carsel, R. F. and R. S. Parrish. 1988. "Developing Joint Probability Distributions of Soil Water Retention Characteristics," *Water Resour. Res.*, 24(5):755–769.

22 Mercer, J. W. and R. M. Cohen. 1990. "A Review of Immiscible Fluids in the Subsurface: Properties, Models, Characterization and Remediation," *J. Contaminant Hydrology*, 6:107–163.

23 Schwille, F. 1967. "Petroleum Contamination of the Subsoil—A Hydrological Problem," in *The Joint Problems of the Oil and Water Industries*, P. Hepple, ed., The Institute of Petroleum, London.

24 Parker, J. C. and R. J. Lenhard. 1989. "Vertical Integration of Three-Phase Flow Equations for Analysis of Light Hydrocarbon Plume Movement," *Transport in Porous Media*, 5:187–206.

Index

T - #0190 - 101024 - C0 - 229/152/11 [13] - CB - 9780877629436 - Gloss Lamination